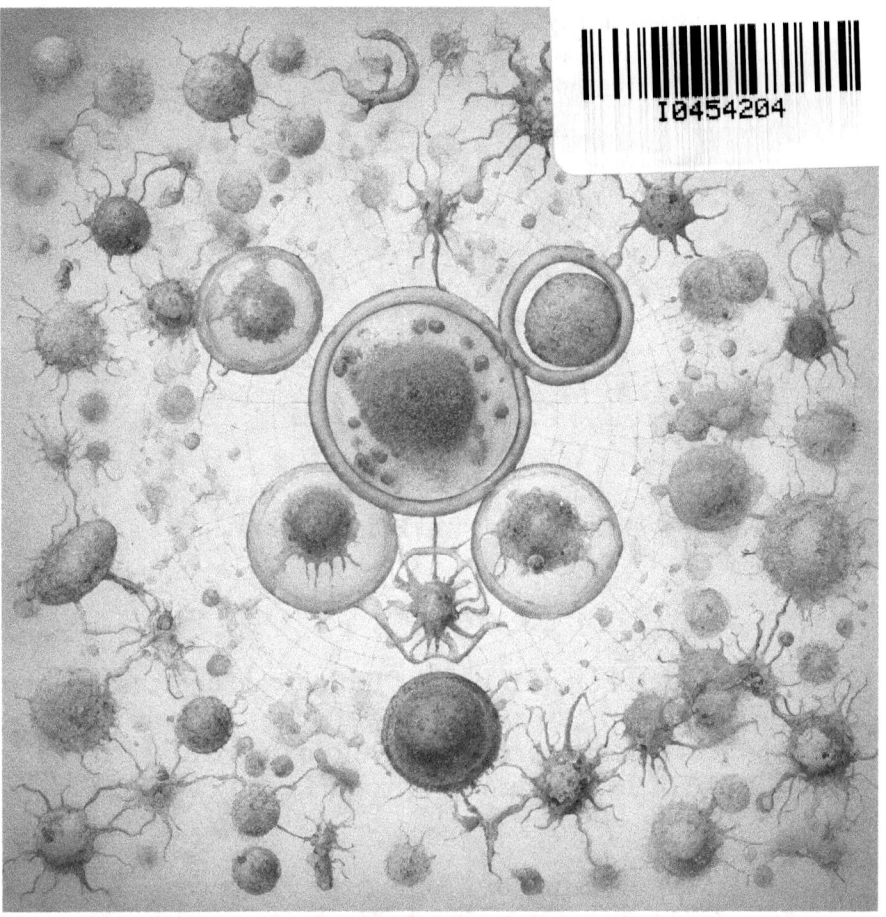

I0454204

Georgii Vasiliadi

Microbiology and Epidemiology

Basics of infectious and invasive diseases development
and possible ways of human infection

Dedicated to the memory of my father
an outstanding scientist
academician, doctor of Biological Sciences

professor of physiology
Georgii Kuzmich Vasiliadi

Microbiology and Epidemiology

Review

**on a textbook of Doctor of Biological Sciences
(Dr.Sci.Biol.), Professor of the Department of Technology
Products Catering of State Technological University
G.K.Vasiliadi "Technology products and catering"**

The work of **Doctor of Biological Sciences**, Professor
G.K.Vasiliadi "Technology products and catering" for masters is
one of the most important areas in biological science. The new
data on microbes classification (Bergey's Manual of Systematic
Bacteriology 2nd Edition, 2001), as well as the list of microbes
and diagnostic testing in accordance with order of Ministry of
Healthcare of the Russian Federation #64 dated 21.02.2000 "On
clinical laboratory tests nomenclature approving" included in
the textbook. The textbook contains the information relevant to
the program, on the basis of which one can have a clear view
of the subject, helps to learn general and special questions of
microbiology, virology, immunology, epidemiology.

Chapters 2 and 5 "Microbiology of microorganisms' natural
habitat" and "Sanitary protection of the country's territory and

infectious diseases spread" are of great interest.

The textbook consists of six chapters fitted out with numerous tables, photographs, diagrams, which undoubtedly allows to improve the quality of learning new material. It is also useful for the students of the preventive medicine faculty and the faculty of pediatrics of medical higher educational institution of the Russian Federation. Despite the great volume of the course content on the topic, the textbook is written in a brief manner and in understandable terms.

The great work of Professor G.K.Vasiliadi deserves high praise.

Taking into account all the mentioned above, I think it possible to recommend the work of Professor G.K.Vasiliadi as a course book to be published and used in higher educational institutions.

Head of department of microbiology, virology and immunology of the SBEI HPE of the North Ossetian State Medical Academy of the Ministry of Healthcare and Social Development of the Russian Federation, Honoured Scientist of the Republic of North Ossetia-Alania, Academician of the Academy of Medical and Technical Sciences, M.D., Professor

L.J.Plachtii

Georgii Vasiliadi

Microbiology and Epidemiology

Basics of infectious and invasive diseases development and possible ways of human infection

Palmarium Academic Publishing

Impressum

Any brand and brand names mentioned in this book are owned by a trademark, brand, or patented and are the brands of the respective rights holders. The use of brand names, product names, brands, product descriptions, common names, etc., even without the exact mentioning in this work, is not the reason for these names to be considered unregistered under any brand and not protected by the brand law and to be used by everyone without any restrictions.

INTRODUCTION

The birth of bacteriology as a separate scientific discipline was preceded by a long period of speculative and then empirical ideas about the character of the phenomena associated with the activity of microorganisms. At the heart of the search, which led to the appearance of bacteriology, have long been three main problems, namely, the causes of infectious diseases, the character of fermentation and decay processes, the source of the smallest living creatures. The results of these problems research formed the foundation for bacteriology arising.

The success of microbiology in the second half of the 19th century allowed the French historian of science P.Tannery to say, "In the face of bacteriological discoveries, the history of other natural sciences in the last decades of the 19th century seems somewhat pale".

Microbiology is a science of the smallest, invisible to the naked eye organisms called microbes or microorganisms. It studies the principles of their life and development, as well as the changes in the human body, animal body, plants and inanimate nature caused by them.

V.Bezhina, a well-known expert on microorganisms studying,

said in one of his speeches, "If any disaster destroyed the humanity on our planet, everything created by man would gradually disappear, but the nature would continue to exist. Rich flora and fauna would fill the entire planet. However, if the disaster touched only bacteria, the plants would gradually droop and fade, then the animals would die, including humans, and the Earth would turn into a barren desert."

The development of microbiology, as well as of other scientific disciplines, stands in close relation to production methods, practical needs, the general progress of science and technology. In accordance with the needs of society, microbiology was divided into general, agricultural, technical (industrial), medical, sanitary and veterinary, and at present time, marine and space microbiology have been formed.

Microbiology, engaged in the study of microorganisms (from the Greek micros – small, bios – life, logos - study), is one of the fundamental biological sciences. Microorganisms take an active part in the circulation of carbon, nitrogen, phosphorus, sulfur in the nature. Thus, in the course of carbon circulation, microorganisms provide constantly a return of

5

carbon dioxide into the atmosphere during the mineralization process of organic material. At the same time, bacteria and fungi play a similar role as photosynthesizing terrestrial plants.

Microorganisms are widely spread in the soil, water and air of all climatic zones. Bacteria are found in water at the depth of 11,000m. They can develop at the pressure of 600 atm and the temperature of 104 -300 ºC.

Many microorganisms live on the surface of the body, in the intestines of humans and animals, on the plants, food and all the objects around us.

BRIEF HISTORY OF MICROBIOLOGY DEVELOPMENT

Long before the discovery of microorganisms the humanity knew some processes caused by their vital activity, e.g.the fermentation of grape juice, milk, dough, etc. In remote times, at the dawn of civilization, the man learned how to make grape wine, koumiss (fermented mare's milk), sour milk and other products using those processes.

The history of microbiology development has the following 5 stages:

Heuristic stage

The biological views of the doctor Empedocles (about 490 – 430 BC) played an important role in spreading the idea of the natural origin of living beings. It was further successfully developed by Democritus (460 – 370 BC) who stated that the world consists of the smallest invisible particles. In the opinion of Democritus, the living beings originated from that period of our world development when the earth, being saturated with moisture, consisted of soft silt. Under the influence of the solar heat the first animals were born in some places. A contemporary of Democritus, the greatest doctor of the ancient world Hippocrates (about 460 – 377 BC) in the books, which composed the so-called Hippocratic Corpus (or Hippocratic Collection), first expressed the hypothesis of wildlife which is a habitat for invisible particles, i.e.miasms. The Hippocratic Corpus contains the most complete set of knowledge and the teachings of Greek physicians in the field of medicine, as well as such theoretical sciences as Anatomy, Physiology and Embryology connected with it. It is valuable enough that this corpus basically represents a materialistic "Democretian" line in Greek philosophy.

6

Hippocrates

Hippocrate's followers finally broke with religion and mysticism in medicine, refusing to explain the origin and nature of disasters by the intervention of supernatural powers. The most important principle of Hippocrate's followers was that it was the patient to be treated, not the disease. Therefore, all the doctor's prescriptions had to be strictly individualized.

An outstanding biologist and researcher of ancient times, whose name came firmly into the history of medicine and biology, doctor Galen (130-200) should also be mentioned. He wrote many works in all branches of medicine. As a great physician, anatomist and physiologist, Galen gained recognition during his lifetime, and his authority in matters of medicine, anatomy and physiology was considered indisputable for one and a half thousand years.

Galen studied the anatomy of sheep, bulls, pigs, dogs, bears and many other vertebrates. He subjected the central and peripheral nervous system to a thorough study. In particular, he investigated the function of the nerves of the spinal cord and tried to determine how they affect breathing and heartbeat.

7

Claudius Galenus (Galen)

At different stages of science development, doctors and natural scientists tried to find out the causes of infectious diseases. In the works of Galen (131-211 BC) and other major scientists of that period, the hypothesis about the animate nature of pathogens of infectious diseases was already expressed.

Avicenna (980-1037 AD) believed that the cause of infectious diseases were the smallest living beings invisible with the naked eye and transmitted through water and air. "The Canon of Medicine", "The book of Healing" by Ibn Sina along with the presentation and commenting of ancient authors, contained the original data and thoughts in the field of medicine and biology. The information on physiology is especially extensive and interesting there. His achievements in the field of medicine are significant. Avicenna is justly considered one of the greatest medical scientists in the history not only of the east, but of the whole world. During his lifetime, he was called the "prince of the doctors" for a good reason. The most famous works of Avicenna on medicine are "The Canon of Medicine" and a poem about

medicine, which was translated into Latin as early as in the 12th century by Gerard of Cremona. Later, this book was translated and published in other European languages.

8

Ibn Sina (Avicenna)

The first to see and describe the microbes was a Dutch scientist Antoni van Leeuwenhoek (1632-1723). He himself made a high quality magnifying glass and designed a microscope with the 160-300 times magnification. In 1678, A. Leeuwenhoek published letters about animalcula viva, that is "live animals" which he observed with the help of his microscope in water, various infusions, feces, and dental plaque. In 1695, he published his work "The Secrets of Nature, revealed by Antoni van Leeuwenhoek". A.Leeuwenhoek was undoubtedly not only the first to discover the microbes, but also sketched them most thoroughly. The discoveries of A.Leeuwenhoek aroused the keen interest among many scientists and served as an impetus for the study of the microworld.

Morphological, physiological and immunological stages

D.S.Samoilovich (1744-1805) suggested that "the plague is called by a special and completely excellent being." He studied and described the infectious disease – the plague. He infused himself some infectious material from a person who had been ill with plague.

The statements made by D.S.Samoilovich regarding the cause of the infectious diseases played an important role in the further development

9

of theoretical and practical issues of plague prevention and many other contagious diseases.

The English doctor E.Jenner (1749-1823) showed that vaccinations of people with the cowpox pathogens protected them from being infected with smallpox.

E.Jenner

Over the course of 25 years, he checked his observations and on May 14, 1795, he conducted a public experiment on the vaccine method. He vaccinated an 8-year-old boy James Phipps with the content of a pustule from the hand of a farmer Sarah Nelme. A month and a half later E.Jenner infused the contents of a smallpox patient's pustules to James. The boy did not get sick.

After repeating this experiment 23 times E.Jenner published an article "Studies of the causes and effects of cowpox".

The method of vaccination proposed by E.Jenner armed medicine with a powerful tool for successful combating this disease. However, E.Jenner's discovery was purely empirical in nature, and its essence remained unclear until the appearance of works by L.Pasteur.

In the second half of the 19th century more sophisticated microscopes appeared, and the microscopy technique was greatly improved. In the study of microorganisms, the attention was paid to biochemical processes, that is, the ability of microbes to ferment organic substances.

The name of the brilliant French scientist, chemist and microbiologist Louis Pasteur (1822-1895) is associated with the most important discoveries in microbiology.

10

L.Pasteur brilliantly confirmed the prediction of the 17th century physicist and chemist R.Boyle that the nature of infectious diseases could be understood by those who could explain the nature of fermentation.

L.Pasteur

L.Pasteur proved the enzymatic nature of alcoholic lactic acid and amylic acid fermentation. He discovered a new (anaerobic) type of respiration of some microbes. He found that the decay is caused by the activity of certain types of microorganisms.

The works of L.Pasteur on the diseases of wine, beer, alkaline worms (pebrine) and the measures of their control are of great importance. The data obtained by him formed the basis for the development of industrial microbiology. The research works of L.Pasteur initiated the use of protective vaccination. He developed vaccines against chicken cholera, malignant anthrax and rabies. L.Pasteur proved that the biogenesis of living beings does not occur. His works served as the basis for the development of medical microbiology.

The discoveries of the German scientist R.Koch (1843-1910), who enriched microbiology with the advanced research methods, are of great importance. Thanks to his improved

11

techniques and methods of microbiological research, R.Koch finally traced the etiology of malignant anthrax, discovered the causative agents of tuberculosis and cholera, extracted tuberculin from tuberculosis microbacteria.

R.Koch

R.Koch made a detailed study of wound infections, developed a method of extracting pathogenic bacteria in pure culture. He established a large school of microbiologists.

The advances in medical microbiology in the sphere of the etiology of infectious diseases led to the need for studying the defensive mechanisms in the body against the infectious agents. The outstanding Russian scientist I.I.Mechnikov (1845-1916) made a great contribution to the development of this very important issue for theory and practice.

The classical works of I.I.Mechnikov, which led to the creation of the theory of phagocytosis, determined a new stage of the medicine development. As a result of the multi-year research, I.I.Mechnikov discovered and studied the process of intracellular digestion in some animals carried out by the cells of mesodermal origin. Those observations suggested that such cells (white blood cells, spleen cells, bone marrow cells, etc.) functioned as a defense against pathogens,

12

penetrated into the body of animals and humans. He named the mesodermal cells phagocytes. The theory of phagocytes was presented by him in 1883 at the 7th congress of Russian natural scientists and doctors in Odessa.

I.I.Mechnikov

The study of phagocytosis was the basis for understanding the essence of inflammation. I.I.Mechnikov showed that the inflammation is an active reaction against pathogenic microbes, ensuting the stability of the organism, which was formed in the process of the evolutionary development of animals and humans. Thus, the beginning of the doctrine of the antagonism of microbes, subsequently used in antibiotics preparation, was made.

The progressive scientific and social activity of I.I.Mechnikov brought on him the prosecution of the tsarist government. The world-renowned scientist in his prime was forced to leave his homeland. He worked in Paris at the Pasteur Institute for 28 years, until the end of his life.

Mechnikov's studying of phagocytosis problems promoted the emergence of a series of works in which it was proved that the specific blood serum substances, i.e.antibodies produced by specific cells under the influence of microbes and their poison, play a major role in the defense reactions.

The German scientist E.Bering and the Japanese researcher S.Kitazato obtained the corresponding immune sera by repeated injections of small doses of tetanus or diphtheria toxin to animals.

13

Those discoveries served as the basis for the production of therapeutic sera against botulism, anaerobic infections, snake venom, etc.

The mechanism of immunity to infectious diseases was studied by the greatest German researcher P.Ehrlich (1854-1915), who created the theory of humoral immunity which gave rise to a long and persistent conflict of opinions and divided the scientists into two camps: the supporters of P.Ehrlich and his opponents led by I.I.Mechnikov.

P.Ehrlich

This controversy caused a rapid flow of research on immunity issues and led to great practical results, namely, more advanced laboratory methods for diagnosing infectious diseases were developed, the vaccines against typhoid fever, cholera, plague and other diseases were obtained.

In 1908, Mechnikov and Ehrlich were awarded the Nobel Prize for their work in the sphere of immunology.

A significant contribution to microbiology was made by L.S.Tsenkovsky (1822-1887). He obtained a highly effective, stable vaccine that has been used in our country for more than 60 years to prevent the malignant anthrax among farm animals.

14

Under the leadership of I.I.Mechnikov, N.F.Gamaleya participated in the creation of the first bacteriological station in Russia and the second Pasteur station in the world. His research work was devoted to the study of infection and immunity, the variability of bacteria, the prevention of typhus, smallpox, plague and other diseases.

The national science laid the foundation for the theory of viruses, the founder of which is considered to be D.I.Ivanovsky (1864-1920). He found that the mosaic tobacco disease, widespread in the Crimea, was caused by a virus having a high degree of infectiousness and strongly showed specificity of action.

D.I.Ivanovsky

The discovery showed that, along with the cellular forms, there were living systems invisible in the conventional light microscopes, passing through fine-pored filters and lacking a cellular structure.

The name of D.K.Zabolotny (1866-1929), who studied the epidemiology of the plague, is surrounded by deserved fame. He proved the therapeutic effect of anti-plague serum, scientifically substantiated the epidemiological role of marmots in the formation of hot spots of plague, specified the lytic effect of serums from syphilis patients on treponema.

15

Molecular genetic stage

The development of microorganism genetics in recent decades should be considered a new stage in microbiology, as a result of which the mechanisms for the exchange of genetic material in bacteria have been elucidated.

The genetics of bacteria and viruses played an important role in the emergence of a new branch of knowledge, i.e. molecular biology. The most important task of molecular biology is the study of the atomic-molecular structure of protein, nucleic acids and other biopolymers, the manifestation of the vital activity of organisms in their simplest elementary form at the molecular level.

Modern medical microbiology has become an extensive scientific discipline. It is subdivided into bacteriology - the science of bacteria – the carriers of a number of infectious diseases; virology - the study of viruses; immunology - the science of the mechanisms of the body protection against pathogenic and non-pathogenic agents; mycology that studies fungi pathogenic to humans; protozoology, the research objects of which are pathogenic single-celled organisms.

Microbiology as an independent field of knowledge has special research methods that allow to solve a number of important issues that are of great importance for health.

Chapter I
General Microbiology

1.1. Morphology, classification, structure and nutrition of microorganisms

Microorganisms are the oldest representatives of living beings on the Earth. They appeared more than 3 billion years ago. The problem of the origin and evolution of microorganisms is extremely complex.

Medical microbiology studies mainly pathogenic bacteria, viruses, fungi, protozoa, conventionally combined under the common name "microbes" or "microorganisms".

In the morphological period of microbiology development, a unilateral (descriptive) approach prevailed in the studies, namely, microbes were studied with disregard for their evolution, speed and connection with the conditions of the external environment.

In 1896, K.Leman and R.Neumann laid the foundation for the creation of a scientifically based classification of microbes, according to which all the organisms

16

were divided into three groups: spherical, rod-shaped and convoluted. However, it was revealed that some features alone were not enough. For this purpose, a complex of features is currently used: phenotypic (morphological, cultural, physiological, and other properties), as well as genotypic (physical and chemical properties of DNA).

It should be pointed out that gene-systematics allows to determine microorganisms not by similarity, but by relationship. In medical microbiology, an international classification has been adopted, presented by D. Bergi. All the microorganisms are combined into the kingdom of prokaryotes, in which two divisions are included: 1 - cyanobacteria, or blue-green algae; 2 - bacterium.

Bacteria (from lat.bacteria - rod) are mostly single-celled organisms devoid of chlorophyll. By their biological properties, they are classified as prokaryotes. The size of the bacteria is measured in micrometers (μm). The sizes of microbes, as well as their forms, are not absolutely constant. Morphological changes are observed in many species of bacteria, that is, they change under the influence of the habitat.

As a rule, such changes are non-hereditary and are called modifications. In appearance, the bacteria are divided into four main forms: spherical (cocci), rod-shaped (bacteria, bacilli and clostridia), convoluted (vibrios, spirilla, spirochetes), filamentous (chlamydobacteria). Fig.1.

Bacteria also include the rod-shaped microorganisms, which, as a rule, do not form a spore. Microbes belong to bacilli and clostridia, most of them reversing spores.

By mutual disposition, the rod-shaped forms are divided into three subgroups: 1) diplobacteria and diplobacilli, located in pairs along the length; 2) streptobacteria and streptobacilli; 3) bacteria and bacilli, which are distributed without a specific system. The crimped forms of bacteria include vibrios, spirilla, and spirochetes.

Long-term studies of microbiologists convincingly show that the individuals of the first generations of microbes, developing in an external environment conducive to their growth, differ from the individuals of subsequent generations.

17

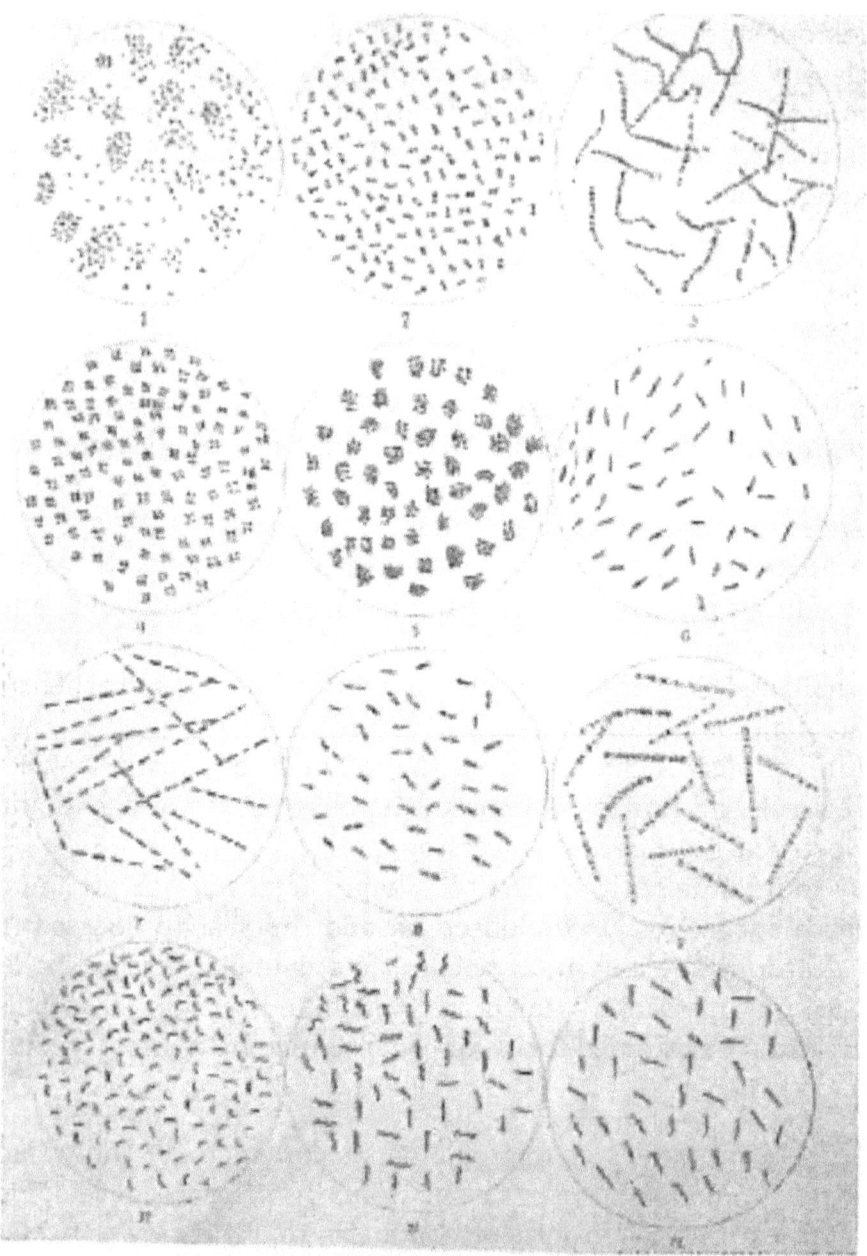

Fig.1. Forms of microorganisms

1-staphylococci   7 – streptobacteria
2 – diplococci   8 - bacilli

3 – streptococci
4 – tetracocci
5 – sarcina
6 – bacteria

9 – streptobacilli
10 – vibrio
11 – spirilla
12 - spirochetes

According to their structure, bacteria are prokaryotes. They differ significantly from the cells - eukaryotes. Prokaryotes usually contain one genome that is not separated by a special membrane from the cytoplasm, do not have mitochondria, but their function is performed by mesosomes, they do not possess amoeboid movement. They

18

consist of a nucleoid, a cytoplasm containing various inclusions, shells and other structures. Fig.2.

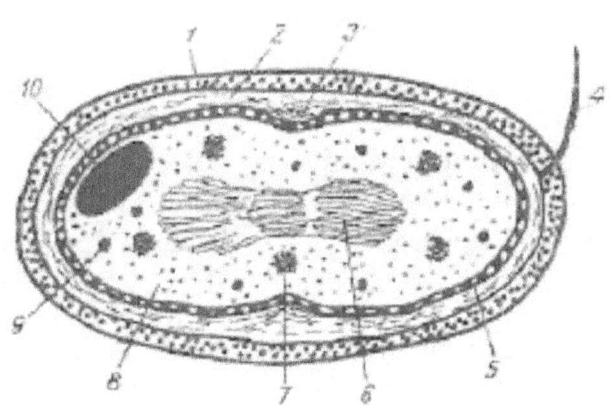

Fig.2. Schematic representation of the

structure of the bacterial cell

1 - capsule; 2 – cell wall; 3 – septum; 4 – flagellum; 5 – plasma membrane; 6 – nucleoid; 7 – ribosomes; 8 – inclusions; 9 – plasmids; 10 – endospore.

The plasma membrane is adjacent to the inner surface of the cell wall. It is a complex highly organized and highly specialized formation. It consists of three layers: lipid, protein and polysaccharide. It performs the function of a partition wall. The active transportation of various substances and ions necessary for the vital activity of the cell is continuously carried out through it with the help of enzymes. In the cell membranes, highly sensitive receptors are localized, with the help of which the cells recognize and process signals from the environment, differentiate nutrients and various antibacterial compounds. The surface of the cytoplasmic membrane contains active enzyme systems that are involved in the synthesis of protein, toxins, enzymes, nucleic acids, and other substances, in oxidative forcing. Such properties of a bacterial cell as osmotic pressure are associated with cytoplasmic membranes, which ensures the constancy of metabolism - homeostasis.

19

The nucleotide (nucleoplasm, carioplasma) of prokaryotes consists of a coil of DNA double strands, it is not delimited from the cytoplasm by any membrane and is not associated with the main protein of the bacterial cell.

Among the bacteria the motile ones, which are subdivided into creeping and floating, are distinguished. The creeping bacteria slowly move (crawl) on the surface as a result of the wavelike contractions of the body, which cause a periodic change in the shape of the cell.

The floating bacteria move freely in a liquid medium. They are equipped with flagella. According to the location of the flagella, motile microbes are divided into four groups (Fig. 3):

Fig.3.Flagella of bacteria

1) monotrichs - bacteria with one flagellum at the end;
2) amphitrichs - bacteria with two polar flagella; 3) lofotrichs — bacteria that have a bundle of flagella at one end; 4) peritrichs - bacteria that possess flagella over the entire surface of the body (Fig. 4).

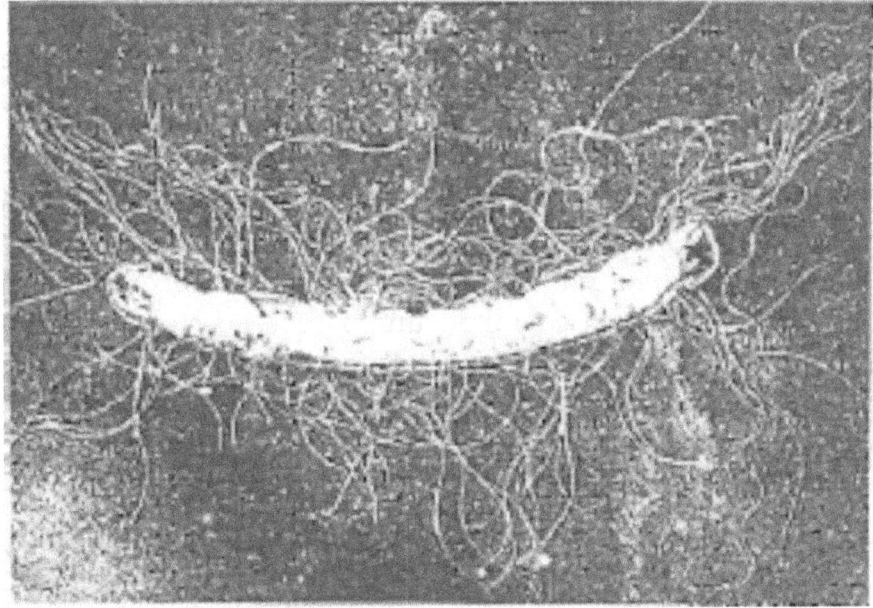

Fig.4.Fragella of Proteus vulgaris in the electron microscope

1.2.Endospores and spore formation. The endospores are

formations of round or oval form. The spore formation is one of the

20

stages of the development cycle of microorganisms, elaborated in the process of evolution in the struggle to preserve the species.

When bacilli enter certain conditions, mainly the unfavorable ones, the structural changes take place in the cell. The spore is approximately 1/10 of the mother cell. This process of the spore formation ends within 18-20 hours. The spores have the increased resistance to the action of the external environment and can last for a long time, 50-100 years, in unfavorable conditions. The spores of some bacilli withstand boiling and the action of high concentrations of disinfectants. They die in the steam chamber under the action of saturated steam at the temperature of 115-125 °C, as well as at the temperature of 150-170 °C in the dry-heat Pasteur oven. Getting into favorable conditions, the spores germinate and turn back into their vegetative forms.

For the designation of microorganisms the double (binary) nomenclature - the name of the genus and species – is adopted. For example, the anthrax bacillus is called Bacillus anthracis. The main (lowest) taxonomic unit is the species. Species - genus - family. A species is a collection of populations having a common origin and genotype, morphological, physiological, and other features that can cause the same processes under certain conditions.

Culture (cells) are microorganisms grown on a nutrient medium.

Strain is a culture of the same species, extracted from different environments and characterized by minor changes in properties.

A clone is a culture of microorganisms extracted from a single cell. In the bacterial environment, there are also bacilli with lateral processes (tuberculosis, paratuberculosis).

1.3. The mechanism of metabolism in microorganisms

All the organisms are characterized by a constant exchange of substances with the surrounding environment. The sources of energy for bacteria are light, inorganic and organic substances.

By the nature of the use of carbon and energy source, the bacteria are divided into four groups:

1.Phototrophs are bacteria, the source of energy for which is light.

2.Chemotrophs use chemicals as the source of energy.

21

3.Autotrophs are bacteria which use carbon dioxide (CO_2) as a source of energy. Some autotrophic bacteria possess the ability to digest polyethylene, boric acid, phenol and other nonorganic substances.

4.Heterotropchs are bacteria which need organic carbon (carbohydrates, fatty acids).

By type of feeding, bacteria are differentiated into:

1) photoautotrophs - the source of energy for them is sunlight, the source of carbon is CO_2;

2) photoheterotrophs - the source of energy is light, the source of carbon is organic compounds;

3) chemoautotrophs - the source of energy is reduced inorganic compounds, the source of carbon is CO_2;

4) chemoheterotrophs - organic compounds are the source of carbon and energy. Most bacteria are in this group.

Saprophytes are growing at the expense of dead substrates. Organic substances are used in the environment. These include most of the types of bacteria that inhabit our planet.

Parasites live on the surface or inside another host organism and nourish at its expense.

A microbial cell uses nutrient substrates to synthesize the constituent parts of its body, enzymes, pigments, growth factors, toxins, deposition of reserve material and the production of energy due to which it exists.

Two opposite and, at the same time, common processes occur in metabolism: the constructive and energy exchange. The constructive metabolism proceeds with the absorption of free energy.

For this type of exchange, a relatively small amount of nutrient material is consumed by the cell. The energy exchange is used to convert energy into a form that is available for its absorption by the cell. The implementation of this process consumes a huge mass of nutrient substrates. Both of these processes are not separate, they are interconnected. The products of incomplete oxidation of the substrate are valuable to the body not only as sources of energy, but also because they are used as integral parts in the construction of the body.

22

The entry of nutrients (molecules) into the cell from the environment and the release of metabolic products determine the interaction of the cell with the environment. The membrane is responsible for the entry of nutrients into the cell. It selectively lets not all the substances through, but only those needed at the moment. Some molecules enter the cell by passive diffusion before leveling off concentrations. Passively, oxygen, water, alcohols and fatty acids pass through the membrane. Water provides the osmotic state of the cell. The passive diffusion is also important for the metabolic products removal from the cell.

The diffusion rate can be increased due to the mechanism of facilitated diffusion, in which the permeases are involved. They are protein-carriers crossing the membrane. They bind to the transported substance, undergo conformational changes, as a result of which the substance is transferred through the membrane to the region of its lower concentration.

The facilitated diffusion accelerates the process of leveling concentrations, but it cannot lead to a concentration of the substance inside the cell. This can be done only by the active transportation of substances through the membrane, carried out by the transmembrane proteins.

The transmembrane proteins are an essential cell adaptation for interaction with the environment. They carry out not only the transfer of molecules through the membrane, but also the active transportatopn, which allows to selectively concentrate the necessary substances inside the cell against the gradient, carrying out a specific chemical reaction. Thus, the active transportation takes place with the help of substances that carry them from a lower concentration to a higher one. The process is accompanied by the expenditure of ATF energy.

1.4. Enzymes and their role in metabolism

Enzymes are biological catalysts of high molecular structure produced by a living cell. They have a protein nature, are strictly specific and play a crucial role in the metabolism of microorganisms.

Microbial enzymes have a variety of action and high activity. They are widely used in industry,

23

agriculture and medicine, and are gradually replacing the enzyme products obtained from higher plants and animals.

Thanks to the use of microbial enzymes in the medical industry, alkaloids, polysaccharides, steroids (hydrocortisone), prednisone, prednisone, etc. are obtained.

Bacteria play an important role in the processing of rubber, cotton, silk, coffee, cocoa. Under their influence the important processes occur, significantly changing these substances in the right direction. Microorganisms have a high synergistic ability.

It has been established that the total mass of bacterial cytoplasm on the Earth significantly exceeds the mass of the cytoplasm of animals. The biochemical activity of microbes has no less general biological significance than photosynthesis. The cessation of the existence of microorganisms would inevitably entail the death of plants and animals.

The enzymes allow some microorganisms to absorb methane, butane and other hydrocarbons and synthesize complex organic compounds from them. Due to the enzymatic ability of yeast cultivated on oil waste (paraffins), protein-vitamin concentrates are used in special industrial-type plants, which are used in animal breeding as a valuable nutrient added to coarse feeds.

Some enzymes are secreted by the cell into the external environment (exoenzymes) for cleavage of complex colloidal alimental material, others are enclosed inside the cell (endoenzymes).
There are constitutive enzymes that are constantly in the cell, regardless of the conditions of its existence and the presence of a catalyzed substrate. These include the main enzymes of cellular metabolism (lipase, carbohydrase, proteinase, oxidase, etc.).
Inductive (adaptive) enzymes are synthesized only when there is

a need for them: they appear only in the presence of a suitable substrate (penicillinase, ammonium decarboxylase, alkaline phosphatase).

The synthesis of the inductive enzymes in microbes occurs due to the presence of free amino acids in the cells and with the participation of the ready-made proteins in bacteria.

24

We have already mentioned the problems of microbial cell nutrition. Note that in the process of feeding the bacteria, passive and active transportation of various substances and ions is distinguished. In case of the passive transfer, the flow of substances moves in accordance with the difference in concentrations or electro-chemical potentials. The active transfer of substances occurs against these gradients due to the energy generated in the cell in the form of "biological sediment". In the regulation of the most important biological processes, the primary role is played by universal cyclic nucleotides, as well as ions of potassium, sodium, calcium, magnesium, which are transferred as a result of the difference in the surface charges of the membrane and the environment surrounding microorganisms.

1.5.Protein metabolism

Microorganisms need different amino acids for their nutrition, growth and life. Some microbes need one amino acid, others need two or more amino acids. Many bacteria have lost the ability to synthesize amino acids. Usually, those species that need vitamins have a need for ready-made amino acids.

The protein content in the cell is about 50%. Different microorganisms have the following protein content (%): 40-80 in bacteria, 40-60 in yeast, 15-40 in fungi.

Some microorganisms synthesize a large number of proteins in the cell - up to 80%. Such microorganisms are considered as possible producers of feed and food protein. The industrial production of such proteins is cost-effective, since microorganisms grow rapidly regardless of the season and weather. As raw materials for their growth the wastes of food and other industries are used. The producers can be bacteria and algae, especially cyanobacteria.

Protein metabolism in bacteria proceeds in two phases. The primary breakdown of the protein to the peptone stage occurs under the influence of the exoprotease secreted by the bacterial cell into the external environment. The secondary decay is due to the action of the endoprotease, which is inherent in all bacteria and is located inside the body of the microbe. The decomposition of the protein to the peptone stage takes place in a nutrient medium with a pH in the range of 7.0 - 8.0.

25

Some microbes produce the enzyme tryptophanase, under the influence of which the indole is formed. Its detection is used in bacteriological diagnostics.
Along with the protein cleavage reaction, the processes of its synthesis also occur. The amino acids are needed to build bacteria proteins.

For the normal development and functioning of a typical bacterial cell, 1000 to 4000 enzymes are needed to ensure the active transport of nutrients into the cell from the environment, regulating the transformation of energy in the cell, carrying out the biosynthesis of amino acids and nucleotides.

The study of amino acid composition showed that the bacterial protein can be considered really native and, therefore, promising as a feed or food product. According to a number of authors (Eroshina, Mavrina, Kuznetsova, 1965), microbacteria grown on hydrocarbon media contain significant amounts of lysine, tryptophan, and methionine, i.e., essential amino acids.

The cell contains 3 - 4% of DNA and 10 - 20% of RNA. In the DNA molecule all the genetic information of the species is encoded. DNA is concentrated mainly in the ribosomes (up to 80%), synthesizing protein, and in the cytoplasm.

1.6.Carbohydrate metabolism

Enzymes that break down carbohydrates, produce hydrolysis of starch to the formation of glucose and maltose. Amylase is contained in many types of microbes: hay, anthrax, diphtheria sticks, cholera vibrio, streptococcus. The presence of this enzyme ensures that the microbe can create a reserve material in the form of polysaccharides in a cell. Some bacteria have a cellulase enzyme that breaks down the fiber.

Under the influence of maltase, sucrase, lactase, the disaccharides ingested into the body undergo hydrolysis and disintegration into monosaccharides, which are then fermented.
Fermentation is characterized by breaking the chain of a carbohydrate molecule and releasing a significant amount of energy.

Enteric-typhoid bacteria in the fermentation of one molecule of glucose, from which then lactic, acetic and formic acids are formed.

26

The manifestation of biological antagonism of fermenting and putrefactive microbes plays an important role in nature and human practice.

1.7.Lipid metabolism

Despite the fact that lipids are not an important component in the nutrient substrate, they have a certain value in the vital activity of bacteria, in particular, they endow microorganisms with high resistance to harmful environmental factors.

Most species of bacteria absorb glycerin, which serves as a source of energy and plastic material for building the constituent structures of a microbe body. The tuberculosis microbacteria and other (acid-resistant) species use glycerin to synthesize lipids. Lipid inclusions in bacterial cells are reserve nutrient material.

When growing yeast and bacteria on mineral media with hydrocarbons,
the more intense synthesis of lipids was observed than with the assimilation of glucose.
The main metabolic processes are carried out with the help of lipase and other lipolytic enzymes strongly associated with the cellular cytoplasm.

Up to 10% of lipids are formed in the cell. Some yeasts and molds synthesize up to 49-60% of lipids. Lipids are part of the cytoplasmic membrane and are part of all other membranes. They can be in the cytoplasm in the form of granules.
Numerous bacteria use methionine as a source of carbohydrates for lipid synthesis.

1.8.Mineral metabolism

For the synthesis of the bacteria body, in addition to nitrogen and carbon, the microorganisms require ash constituents (sulfur, phosphorus, potassium, calcium) and trace elements (boron, molybdenum, zinc, manganese, cobalt, nickel, copper, iodine, bromine, etc.).

One of the most important elements that make up the bacterial cytoplasm is sulfur, which takes part in synthetic reactions.

27

Phosphorus is contained in nucleic acids, numerous enzymes, various phospholipids and other organic compounds in the form of P_2O_5. During the oxidative processes, the energy accumulated in the cytoplasm of microbial cells is released. The adenazine triphosphoric acid (ATP) and adenazindiphosphoric acid (ADP) play a major role in the energy metabolism of microbial cells. Phosphorus is a part of the most important compounds of the cell cytoplasm. The phosphorus content in the form of P_2O_5 in the dry matter of bacterial cells makes 5%.

Microelements are involved in the synthesis of the enzymes that activate them. Molybdenum and boron are essential substances for nitrogen-fixing bacteria.

1.9.Bacteria breathing

As you know, the atmospheric air contains approximately 78% of nitrogen, 20% of oxygen and 0,03 – 0,09% of carbon dioxide. The nitrogen gas can only be used by nitrogen-fixing bacteria, the carbon dioxide — the only carbon source — is utilized by the autotrophic bacteria. Oxygen plays a very important role in the metabolism of most species of bacteria, in respiration and energy production.

The bacteria breathing is a complex process, which is accompanied by the release of energy required by microorganisms for the synthesis of various organic compounds. The bacteria, like higher animals and plants, use oxygen for respiration. L. Pasteur found out that the energy required for the vital activity of certain types of bacteria was obtained in the fermentation process.

All the respiratory bacteria are divided into obligate aerobes, microaerophiles, facultative anaerobes, obligate anaerobes.

The obligate aerobes, developing in the presence of 20% of oxygen in the atmosphere, grow on the surface of liquid or dense nutrient media, contain enzymes with which the hydrogen is transferred from the oxidized substrate to the oxygen of the air.

The microaerophiles need significantly less oxygen. A high concentration of oxygen, while not killing bacteria, delays their growth.

The facultative anaerobes can multiply both in the presence and in the absence of molecular oxygen.

28

The obligate anaerobes are bacteria for which the presence of molecular oxygen is a harmful, growth retarding factor.

Under anaerobic conditions, the yeasts obtain the energy necessary for vital activity by fermenting mono and disaccharides.

$$C_6H_{12}O_6 - 2CH_3CH_2OH + 2CO_2 + 118 \text{ kJ}$$

The aerobic bacteria in the process of respiration oxidize various organic substances (carbohydrates, proteins, fats, alcohols, organic acids and other compounds). With full oxidation, a certain amount of calories of heat is released, which corresponds to the stock of potential energy that has been accumulated in the carbohydrate molecule during its photosynthesis in green plants from carbon dioxide and water. Not all aerobes complete the oxidation reaction. With the excess of carbohydrates in the medium, the incomplete oxidation products are formed, which contain energy. The final products of incomplete aerobic oxidation of sugar can be organic acids: citric, malic, oxalic, succinic. They are formed by mold fungi. Aerobic respiration also takes place with acetic acid bacteria, which produce acetic acid and water during the oxidation of ethyl alcohol.

$$CH_3CH_2OH + O_2 - CH_3COOH + H_2O + 494,4 \text{ kJ}$$

The oxidation of ethyl alcohol with the acetic acid bacteria can go even further - until carbon and water appear with the formation of large amounts of energy.

$$CH_3CH_2OH + 3O_2 - 2CO_2 + 3H_2O + 1366 \text{ kJ}$$

During anaerobic respiration, the oxidized inorganic compounds easily give off the oxygen and turn into more reduced forms. The recovery of nitrates to molecular nitrogen.

$$5C_6H_{12}O_6 + 24KNO_3 --- 24KHCO_3 + 18H_2O + 126CO_2 + 1760 \text{ kJ}$$

The anaerobic process in microbes was first established in 1861 by L. Pasteur.

29

The anaerobic bacteria ferment mainly nitrogen-free compounds, causing the fermentation phenomena. However, there is no clear line between aerobic and anaerobic types of breathing. For example, yeast can change the anaerobic type of respiration to the aerobic one. First, they split sugar to form alcohol and carbon dioxide, and with increased aeration, glucose is split into water and carbon dioxide. The breath of bacteria occurs with the participation of enzymes such as oxidases and dehydrases, which have a pronounced specificity and diversity of action.

The intensity of the processes of aerobic respiration depends on the age of the culture, temperature and substrates. The actively growing cultures consume 2500 - 5000 mm2 of oxygen per 1 mg of dry matter of bacteria per hour, while those starving or completely deprived of nitrogenous nutrition - only 10-150mm2.

The exaggerated breathing and accelerated metabolism are associated with the speed of cell multiplication, with increased protein synthesis in the cell, which leads to the increase in the reducing properties of the medium in which microbes develop. Thus, the essence of energy metabolism consists in obtaining energy produced in the process of direct biological oxidation of substances by atmospheric oxygen or by dehydrogenation (withdrawal of a hydrogen electron from a substrate).

Consequently, the bacterium breathing processes are very complex and represent a long chain of successive reduction-oxidation reactions involving many enzymatic systems that transfer electrons from the system with the highest negative potential to the system with the highest positive potential.

It is established that yeast also has the ability to change its type of respiration depending on the presence or absence of oxygen.

The toxic effect of oxygen in relation to anaerobes is explained by the fact that in the presence of oxygen hydrogen peroxide is formed. Anaerobes do not have the ability to produce catalase. It is not the oxygen which is poisonous, it is only the hydrogen peroxide.

1.10. Microbial pigment formation
Some types of bacteria and fungi that inhabit the soil, water and air, have the ability to produce pigments.

30

The colonies of pigment-forming microbes on dense media are coloured in red (Serratia marcescens, actinomycetes, yeast), pink (pink micrococcus), golden (Staphylococcus aureus).
Sarcinas colonies are coloured in yellow, lemon, golden color.

Some microorganisms produce two pigments or more.

The formation of pigments occurs with good access of oxygen, the temperature of 20-25 °C, and in most species with the diffused sunlight.
The pigments are divided into those soluble in water and alcohol and those insoluble in water and alcohol.

The pigment formation in microbes has a definite physiological significance. It is possible that pigments perform the function of an acceptor of hydrogen in the process of respiration, provide protection against the natural ultraviolet reaction, participate in the synthesis reactions, and also have an antibiotic effect.

1.11. Luminescence of microorganisms

Glow (luminescence) is a peculiar form of energy release during the oxidative processes. The glow is the more intense, the stronger the flow of oxygen to the bacteria is.

The glow of meat, fish scales and other objects was noted by Aristotle (384-322 BC). Luminous microorganisms sometimes penetrate into the body and muscles of small crustaceans, causing a bright glow of these animals at night on the seashore. The glowing termites, ants and spiders were found. It is believed that the luminous bacteria are the source of light. Some fish have developed special organs to hold luminous bacteria as symbionts, which serve as the source of light. Some mushrooms that live in old tree stumps and roots also glow.

At the beginning of the 20th century, H. Molish proposed the use of luminous bacteria for the "safe lamps" used in powder magazine. These include a large group of physiologically similar, but morphologically different bacteria (cocci, rods, vibrios). Most species of luminous bacteria are extracted from seawater. They do not cause rotting, although they grow well on fish and meat substrates, and are cultivated in the ordinary media.

31

A typical representative of photogenic microbes is Photobacterium phosphoreum - a fixed coccoid bacterium that does not dilute gelatin, which develops at 28 °C. In the group of photogenic bacteria, the species that are pathogenic for humans are not determined.

1.12.Aromatic microbes.

Some microorganisms produce volatile aromatic substances, for example, acetic, ethyl and acetic-amyl ethers, which give aromatic properties to wine, beer, lactic acid products, hay, etc. This group includes yeast, lactic acid bacteria, mold fungi, actinomycetes. Many fragrances used in the creation of flavors, give the finished product not only the taste but also the aroma. The antioxidant properties of fragrant substances that make up the essential oils of fennel, laurel, coriander are well known. The mustard essential oils have antimicrobial properties. Flavoring agents consist of or are made from the components of the following categories: flavoring substances, flavoring preparations, thermal, technological, smoke flavors.

1.13.Reproduction and growth of bacteria

By the reproduction of bacteria their ability to reproduce themselves, increasing the number of individuals per unit volume is imply. Rum means an increase in the mass of bacteria as a result of the synthesis of cellular material. The bacteria multiply by simple division – the transverse one with the formation of diverse combinations of cells. Figure 5.

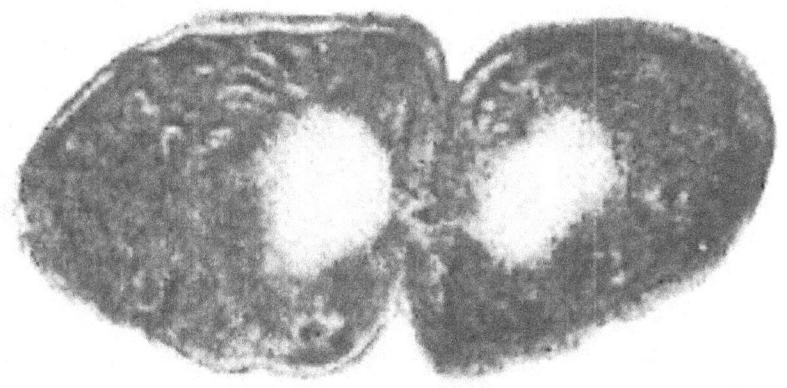

Fig.5 Ultrathin section of the dividing cell E.coli

32

In the process of amitotic binary division of bacteria, an important condition is the replication (doubling) of DNA. Herewith, the breaking of hydrogen bonds and the synthesis of two DNA chains, each of which is in the daughter cells, occurs. Subsequently, single-stranded DNAs are joined by hydrogen bonds and double-stranded DNAs are again formed to carry out the genetic information. The DNA replication and cell division occurs at a certain rate inherent in each species. After a certain number of generations, the cells age and die.

The reproduction of microbes occurs though quickly, but not infinitely. Thus, according to the calculations of Stent (a), one E. coli in the active growth phase in 24 hours of cultivation with the cell division after 20 minutes would give the offspring, the weight of which would be about 10 thousand tons. However, as it is well known, the unlimited reproduction of microbes does not occur.

The separation of bacteria, in its turn, occurs in three ways: 1) the breaking apart the division, when two individual cells, repeatedly breaking over at the junction, tear the cytoplasmic bridge and repel each other; at the same time the chains are formed (anthrax bacilli); 2) the splitting separation, in which, after dividing, the cells isolate and one of them splits on the surface of the other (separate forms of Escherichia, Fig.6); 3) the cross-sectional separation, when one of the divided cells is the point of its contact with another cell, forming the Roman five or cuneiform (Corynebacterium diphtheria).

Most bacteria species reproduce asexually (isomorphic division). However, a reduced sexual process, carried out between males and females is revealed in the bacteria.
The rate of bacteria reproduction in the population is different. It depends on the microbe, the age of the culture, the nutrient medium, the temperature, the carbon dioxide concentration and many other factors.

The duration of generation under favorable conditions in Clostridium perfringens, Streptococcus faecalis is 15 minutes, while for mammalian tissue culture cells - 1 day. Therefore, the bacteria multiply almost 100 times faster than tissue culture cells.
The reproduction of bacteria occurs according to certain laws. The reproduction of the culture of microbes on an irremovable medium is uneven. At the same time, several stages are revealed, according to Piatkin. Fig.6.

33

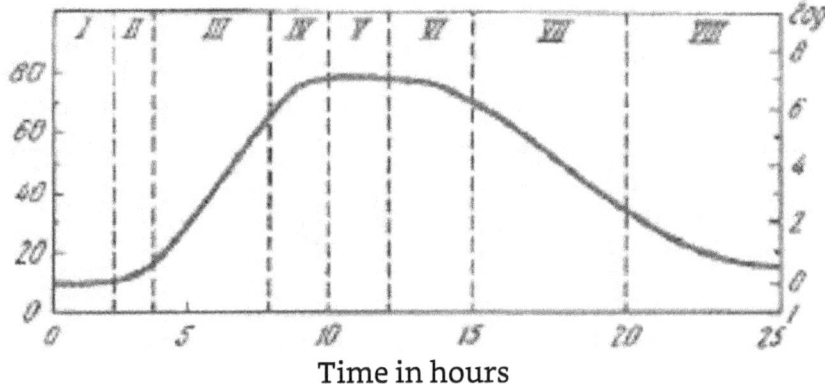

Time in hours

Fig.6. Bacterial growth curve

1. The initial stationary phase is the time from the moment of bacteria inoculation to the beginning of their growth. In this phase, the number of live bacteria may even decrease. Its duration is 1-2 hours.

2. The reproduction delay phase is characterized by the increase in the rate of increase in the size of bacteria and weak growth.

3. The exponential (logarithmic) phase is distinguished by the fact that the logarithm of the number of cells increases linearly, depending on time. Cells are divided at a maximum constant speed. In this phase, the bacteria have the highest biochemical and biological activity. The duration of this phase is 5-6 hours.

4. The negative acceleration phase, during which growth rate of bacteria ceases to be maximum. The number of dividing individuals decreases. This phase takes about 2 hours.

5. The stationary phase of the maximum, when the number of new bacteria is almost equal to the number of the dead ones. It lasts 2 hours.

6. The phase of death acceleration, during which there is an imbalance between the stationary phase and the speed of the bacteria death. It lasts 3 hours.

7. The phase of logarithmic death, when the death of individuals occurs at a constant rate. It lasts about 5 hours.

8. The phase of the extinction rate decreasing – the survived

individuals get into the state of rest.

The duration of separate phases is given conditionally, since it can vary depending on the type of bacteria and cultivation conditions. For example, E. coli is divided every 20-30 minutes. By the way, the number of Escherichia coli among other representatives of the intestinal microflora does not

34

exceed 1%, but they play an important role in the functioning of the gastrointestinal tract.

The intestinal bacterium E. coli is the main competitor of opportunistic microflora in the colonization of intestines. E. coli produces a number of vitamins necessary for humans: B1, B2, B3, B4, B12, K. It participates in the exchange of cholesterol, bilirubin and fatty acids.
E. coli appears in the human intestine in the first days after birth and is preserved throughout life. Fig. 7.

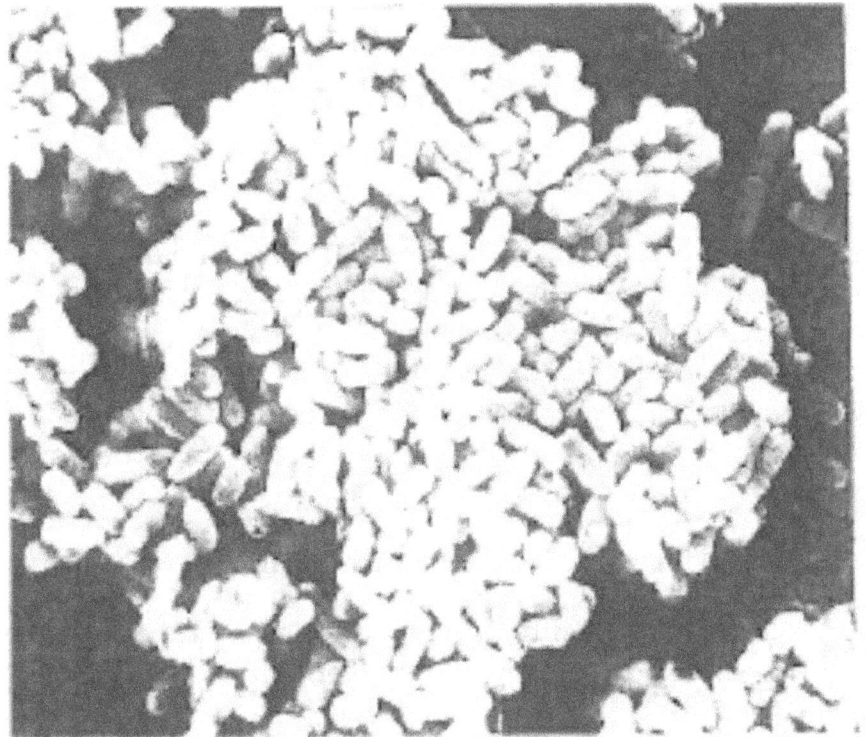

Fig.7.Cells of Escherichia coli (intestinal bacterium)

The typhoid salmonella - 23 minutes, the pathogenic streptococci - 30 minutes.

In addition to reproduction and growth, the bacteria have the age variability, i.e.the ability to change individuals in different stages of growth, maturation and ageing. These changes are observed in the normal cycle of individual bacterial development.

It should be noted that one of the possible reasons that slow down the bacteria growth is the depletion of the nutrient medium. However, this is a rather complicated process that cannot be explained only by the shortage of power sources in the environment. The disappearance from the environment of substances specific to this bacterial species is of great importance. For example, after the cultivation of Escherichia coli or

35

or Staphylococcus aureus in meat broth the bacteria inoculation of the same species on this broth is not accompanied by the growth, but the growth is observed if other microbes are inoculated on this depleted medium.

The knowledge of the patterns of development is of practical importance when growing and preserving the crops on liquid and dense nutrient media.

1.14. Basic principles of bacteria cultivation

Under laboratory conditions, the bacteria are grown on the nutrient media. The temperature conditions are of great importance for the growth and reproduction of bacteria. All microorganisms, in relation to the temperature regime, are divided into three groups: psychrophilic (cold-loving), mesophilic (medium), thermophilic (thermophilic) (tab.1). The bacteria can multiply in a wide range of temperatures - from 0 °C to + 90 °C.

For the life of bacteria pH is of great importance. In the process of evolution, each type of microbes has adapted itself to exist within certain pH limits, beyond which its vital activity is impossible.

It is believed that pH affects the activity of enzymes. Depending on the pH, the lactic acid in the acidic medium is in the form of molecules, and in alkaline - in the form of ions. The saprophytes can live in conditions with an extremely wide pH range - from 2 to 8.5. The pathogenic species of microbes grow at pH 6-8 (Tab.1).

Table 1. Differentiation of microbes in relation to the temperature regime.

Microbial group	Temperature limits of microbial reproduction, °C	Habitat
Psychrophilic	0 – 20	Reservoirs of cold seas and oceans, the soils of polar countries and the permafrost zone
Mesophilic	20 – 45	
Thermophilic	45 - 70	Animal and human organism

36

The nutrient media must contain everything necessary for the growth and development of the microbes: nitrogen, carbon, inorganic compounds in the form of salts, vitamins, trace elements and a certain pH.

The nutrient media are divided into four main groups: universal, special, selective (elective) and differential-diagnostic.

1. The universal media (meat-peptone broth, meat-peptone agar) contains the nutrients, in the presence of which many types of pathogenic and non-pathogenic bacteria grow.

2.The special media are used to grow bacteria that do not multiply on universal media. The special media include blood agar, whey agar, whey broth, etc.

3. The electoral (elective) media: only certain types of bacteria develop well in them, other types do not grow well or at all. These include the enrichment environments in which the species of interest for a scientist grows more intensively and faster than the accompanying bacteria.

4. The differential diagnostic media are used to differentiate certain types of bacteria according to their cultural and biochemical properties.

Currently, many nutrient media are manufactured at the factories and produced in a powder form. They are easy to use, stable and very effective.

The consistency of the nutrient medium are dense (meat-peptone agar, meat-peptone gelatin, folded serum, potatoes, folded egg white), semi-liquid (meat-peptone agar) and liquid (peptone water, meat-peptone broth, sugar broth, etc.).

On dense nutrient media, the microbes form colonies of various shapes and sizes, which are the visible accumulations of individuals of one type of microorganisms, which are formed as a result of reproduction from one or several cells. The colonies are flat, convex, dome-shaped, indented, their surface is smooth (S-shape), rough (R-shape), striated, uneven, the edges are even, notched, fibrous.

The colonies differ in consistency, density, and color. They are transparent and opaque, colored and colorless, moist, dry and slimy.

37

1.15. The influence of environmental factors on microorganisms

The prokaryotes are characterized by the ability to exist in a much wider range of environmental conditions is characteristic than the eukaryotic organisms. Among prokaryotes, there are the organisms capable of existing in underwater volcanic sources (at the temperature up to 30 °C) acidic at pH 1 and below and alkaline at pH 11 and above media at the pressure of 1000 atmospheres.

The microbes tolerate low temperatures relatively easily. The vibrio cholerae does not lose its viability from the temperature of -32 °C: some types of bacteria remain revivable at the temperatures of liquid air (-290 °C), liquid hydrogen (-253 °C). The corynebacteria diphtheria tolerate freezing for 3 months. The typhoid salmonella survives in ice for a long time. The bacillus spores maintain the temperature of -250 °C for three days. Many microorganisms are resistant to low temperatures. For example, the Japanese encephalitis virus in 10% of the suspended matter of the brain does not reduce its pathogenicity at –70 °C for 1 year, the causative agents of influenza at –70 °C for up to 6 months.

Each of us has repeatedly seen that low temperatures suspend putrefactive and fermenting processes. This principle is based on the use in practice of glaciers, cellars and refrigeration units for the preservation of food.

But there are very sensitive pathogenic microbes to low temperatures (meningococci, gonococci). Taking this data into account, the materials used for meningitis and gonorrhea are delivered to the laboratory protected from cooling. At low temperatures, the metabolism slows down.

The alternating high and low temperatures make a detrimental effect on microbes. Most asporogenic bacteria die at 58-60 °C within 30-60 minutes. The spores of bacilli and clostridia are more resistant than vegetative forms. They withstand boiling from several minutes to 3 hours. They die from the dry heat at 160-170 °C during 1 hour. The heating at 120 °C under a vapor pressure of 2 atm kills them within 20-30 minutes. The basis of the bactericidal action of high temperatures is the damage of ribosomes, the denaturation of proteins and the violation of the osmotic barrier. High temperatures cause the destruction of viruses rather quickly. Hepatitis A, poliomyelitis and other viruses persist.

38

in water, in feces of patients or carriers, resistant to heating at a temperature of 60 ° C and the action of small concentrations of chlorine in water.

Microorganisms have different resistance to drying, to which gonococci, meningococci, treponemes, leptospira, hemophilic bacteria, and phages are sensitive. The cholera vibrio does not die under the influence of drying for 2 days, shegella - 7, plague bacillus - 8, diphtheria - 30, typhoid - 70, staphylococcus and mycobacterium tuberculosis - 90 days. The dried sputum of tuberculosis patients remains infectious for 10 months. The anthrax bacillus spores persist for up to 100 years, mold fungi last for 20 years.

Drying is accompanied by dehydration of the cytoplasm and denaturation of bacteria proteins.

One of the methods of food preservation is sublimation - dehydration at a low temperature and high vacuum and is accompanied by evaporation of water, rapid cooling and freezing. The ice formed in the product easily sublimates bypassing the liquid phase. The duration of food preservation is more than 2 years. Freeze dehydration ensures the preservation of all sugars, vitamins, enzymes and other components. Drying in vacuum at a low temperature does not kill bacteria and viruses. This crop preservation method is used in the production of stable and long-lasting storage of live vaccines against tuberculosis, plague, tularemia, brucellosis, smallpox, influenza and other diseases.

Some bacteria tolerate the action of light relatively easily. The sunlight has a harmful effect on others. The most bactericidal is the direct sunlight.

Different types of radiation have a bactericidal or sterilizing effect. These include ultraviolet rays (electromagnetic rays with a wavelength of 200-300 nm), x-rays (electromagnetic rays with a wavelength of 0.005-2 nm), gamma rays (short-wave x-rays).

The experiments on the use of shortwave rays for disinfecting chambers, disinfecting infectious material, preserving food, preparing vaccines, treating operative rooms, delivery rooms, etc. have shown that they have very high bactericidal action.

39

The viruses are less resistant to X-rays than bacteria. The viruses pathogenic for animals are inactivated by radiation of 44000 - 280000r.

The ionizing radiation can be used in the practice of food sterilization. This method of cold sterilization has several advantages: it does not change the quality of products due to the denaturation of its constituent parts (proteins, polysaccharides, vitamins), which occurs during heat sterilization.

The bacteria easily tolerate high atmospheric pressure, they do not change noticeably from the pressure of 100-900 atm at the depth of 1000-10000m of the seas and oceans. The yeast retains its vital activity at the pressure of 500 atm. Some bacteria, yeast, mold can withstand the pressure of 3000, photo-pathogenic viruses - 5000 atm.

Ultrasound has waves of bactericidal properties (the waves with the frequency of about 20,000 Hz), which is currently used to sterilize food products, manufacture vaccines and disinfect objects. The mechanism of the bactericidal action of ultrasound is that in the cytoplasm of bacteria in an aqueous medium, a cavitation pocket is formed, which is filled with liquid vapors. The pressure of up to 10,000 atm arises in the bulb, which leads to disintegration of cytoplasmic structures.

By their action on bacteria, bactericidal chemicals are divided into surfactants, phenols and their derivatives, crystals, salts of heavy metals, oxidizers, and a group of formaldehyde. The surfactants can accumulate at the interface and cause a sharp decrease in surface tension, which leads to disruption of the normal functioning of the cell wall and the cytoplasmic membrane. These include fatty acids, including soaps, which cause damage only to the cell wall and do not penetrate into the cell.

Phenol, cresol and their derivatives initially damage the cell wall, and then the cell proteins. Brilliant green, rivanol, trypaflavin, acriflavin, etc. are referred to dyes with bactericidal properties.

Heavy metal salts (lead, copper, zinc, silver, mercury) cause coagulation of cell proteins. A number of metal has oligodynamic action. For example, silverware, silver-plated items, silver-plated sand when contacting with water transfers it its bactericidal

40

properties against many types of bacteria. The mechanism - positively charged silver ions are adsorbed by the negatively charged surface of bacteria and alter the permeability of their cytoplasmic membrane - the violation of nutrition and growth. Also, the viruses are inactivated.

The oxidants act on sulfhydryl groups of active proteins. These include chlorine, which affects dehydrases, hydrolases, amylases, and bacterial proteases, strokes used in water decontamination, bleach and chloramine, used for disinfection purposes.

Iodine in the form of the alcoholic solution oxidizes

the active groups of the bacteria cytoplasm proteins. Potassium permanganate and hydrogen peroxide have oxidizing properties. Almost all the viruses are stored in glycerin solution. The viruses are destroyed by caustic soda, caustic potassium, chloramine, bleach.

Formaldehyde is used in the form of the 40% solution - the so-called formalin. Its antimicrobial effect is believed to be due to the fact that it joins the amino groups of proteins and causes their denaturation. Formaldehyde kills both the vegetative forms and the spores. It is used to neutralize diphtheria and tetanus toxins, in the course of which they turn into toxoids.

Under natural conditions, the microorganisms are part of a biocenosis (a set of plants and animals that inhabit a part of the habitat with more or less homogeneous living conditions).

Microbes are found in nature in associations, which struggle constantly for existence. Certain species that have adapted to this environment have more pronounced antagonistic properties with respect to other species entering the new habitat. The lactic acid bacteria have antagonistic properties against pathogens of dysentery, plague, etc. The pseudomonas bacterium inhibits the growth of shigella, salmonella, anthrax bacilli, and cholera species.

Normal inhabitants of the human body have especially powerful antagonistic properties: E.coli, Str.faccalis, lactic acid bacteria, actinomycetes, the microflora of the skin, nasopharynx, etc.

41

Under certain conditions of microbial existence, antagonistic relationships result from the lack of nutrients, and then some microbes are forced to feed at the expense of others.

Antagonistic relationships are also established among the viruses, when one virus protects the body from the introduction of another virus into it. In virology, this phenomenon is called viral interference.

There are several types of relationships between different groups of microbes: symbiosis, metabiosis, satelliteism, synergism, antagonism.

Symbiosis is the cohabitation of organisms of different species, usually bringing them mutual benefit. They develop better together than each of them individually.

Metabiosis is a type of relationship when one organism continues the process caused by another, freeing it from the products of vital activity and thus creating conditions for its further development.

With satelliteism, one of the cohabitants, called the favoring microbe, stimulates the growth of the other articulate (some yeasts and sarcins, which produce amino acids, vitamins and other substances, promote the growth of more nutrient-intensive microbes).

Synergism is characterized by the effort of physiological functions in members of the microbial association (yeast and lactic acid bacteria, fuzobakterii and borrelia).

In antagonistic relationships, there is a struggle for oxygen, nutrients and habitat. Bacteria, fungi, higher plants produce substances called antibiotics, which have a detrimental effect on other microbes. They are widely used in the treatment of many infectious diseases.

In the neutralization of the environment from pathogenic microorganisms, due to antagonism, a great role is played by phages that are widespread in soil and water, and phytoncides are volatile substances of many plants.

Chapter II
Microbiology of the natural habitat of microorganisms

Microbes are ubiquitous in our environment. They are found in soil, water, air, on plants, food, objects,

42

in humans and animals. Unlike plants and animals, the microorganisms can use methane, hydrogen in their metabolism, molecular nitrogen, carbon monoxide and turn them into compounds that are absorbed by plants and animals. So, for example, the putrefactive microbes, splitting dead plants and animal corpses, return to the atmosphere 90% of CO_2 absorbed by the plants.

For many species of microorganisms, the intestine is a biotope, that is, the only natural habitat for them. Therefore, the detection of intestinal microflora in the material under study (foodstuff) serves as direct evidence of fecal contamination of the object and indicates the possible presence of intestinal pathogens in it (typhoid, cholera, etc.). Microorganisms secreted in these cases appear in the genus of indicators of health problems.

2.1. Soil microflora

Soil fertility depends not only on the presence of inorganic organic substances, but also on various types of microorganisms causing the qualitative composition of the soil. The number of microbes in 1 g of soil reaches great sizes: from 200 million bacteria in clay soil to 5 billion in black earth. 1 g of arable soil layer contains 1-10 billion bacteria. Especially important are nitrogen-fixing bactecria. Blue-green algae play an important role in the enrichment of soil with nitrogen. In the arable layer of cultivated soil on an area of 1 hectare 5-6 tons of microbial mass can be contained.

Soil contamination by microorganisms is closely dependent on the degree of contamination by fecal matter and urine.
The soil is an unfavorable environment for most pathogenic species of bacteria, viruses, fungi. the simplest. However, the soil as a factor in the transmission of a number of pathogens of infectious diseases is a very complex substrate.

The spores of anthrax bacilli, clostridium tetanus, anaerobic infections, botulism, and many soil microbes persist for long in the soil.
The soil serves as a habitat for various animals (rodents), on which the carriers of pathogens of plague, tularemia, mosquito fever, hemorrhagic fevers, encephalitis, and rural leishmaniasis parasitize.

43

etc. Particularly great is the role of the soil in the transmission of decay invasions (roundworm, whipworm, hookworm, etc.).

Considering the specific epidemiological role of the soil as a factor in the spread of certain infectious diseases of animals and humans, a number of measures are taken in sanitary and epidemic practice aimed at protecting the soil from contamination and infection with pathogenic microorganism species.

2.2. Water microflora

The water of the open sea and freshwater bodies, as well as the soil, is the natural habitat of a variety of bacteria, fungi, viruses, protozoa. The groundwater contains single microorganisms. The microflora of river water depends on the degree of their biological pollution and the quality of wastewater treatment discharged into river beds. Microorganisms are also widely distributed in the waters of the seas and oceans. They were found at the depth of over 10,000m.

The degree of water contamination by organisms is usually expressed by saprobity, by which the collection of living creatures living in waters containing large concentrations of animal or plant debris is meant.

Depending on the degree of pollution, the pathogenic bacteria can be kept in water bodies for a certain time and remain viable. So, for example, the salmonella in tap, river and well water may stay from 2 days to 3 months, shigella - 5-9 days, leptospira - 7-150 days. The vibrio cholerae survive in the water of rivers, seas up to several months, the causative agent of tularemia - from several days to 3 months.

The tapwater is considered good if the total number of microbes in 1 ml is 100, doubtful - with 100-150 microbes, contaminated - with 500 and more. In the water of wells and open water bodies, the number of microbes in 1 ml should not exceed 1000.

The degree of biological contamination of water is assessed by coli index and coli titer. Coli-index is the number of E. coli individuals found in 1 liter of water. The tapwater is considered good if its index is within 2-3, and the coli-titer is 300.

Water is a very powerful factor in the transmission of a number of infectious diseases: typhoid fever, salmonella gastroenteritis, cholera,

44

dysentery, leptospirosis, etc. In connection with the large sanitary and epidemiological role of water in relation to the intestinal group of diseases, it became necessary to develop accelerated methods for the detection of E. coli and pathogenic bacteria in water.

For a more complete and in-depth study of the microflora of soil and water, capillary microscopy is used. It lies in the fact that very thin capillaries are placed in the soil or water bodies, the contents of which are then subjected to bacteriological examination. This method allows the identification of microorganisms that do not grow on ordinary nutrient media and have remained unknown to microbiologists for many years.

2.3. Air microflora

The composition of microbes in the air depends on the degree of pollution with mineral and mineral suspensions, temperature, precipitation, the nature of the terrain, humidity and other factors. The higher the concentration in the air of dust, smoke, soot is, the more microbes there are.

Over the surface of mountains, seas, arctic countries, oceans, the microbes are rare.

The microflora of air consists of a wide variety of species that come into it from the soil, plants and living organisms.

The number of microbes in the air varies in wide ranges - from several copies to many tens of thousands in 1m3. So, for example, the air of the Arctic contains 2-3 microbes per 20m3, while in the industrial cities, 1ml of air shows a huge amount of bacteria. In the forest, especially coniferous, there are very few microbes; the volatile substances of plants, i.e the phytoncides with bactericidal properties, have a destructive effect on them. Above Moscow, at the altitude of 500m, 1,100-2,700 microbes were found in 1m3 of air, while at the altitude of 2000m - from 50 to 700. Spore-bearing microbes and mold fungi were found at the altitude of 20km. Some microorganisms were found at the altitude of 61-77km. 1 g of dust contains up to 1 million bacteria.

Depending on the season of the year, the composition and amount of microflora change in the air. If we take the total number of microbes in winter as 1, then in spring it will be 1.7, in summer - 2, in autumn - 1.2.

45

For indoor air, the sanitary indicative microorganisms are staphylococci, green streptococci, and hemolytic streptococci and staphylococci are indicators of direct epidemiological danger.

When sneezing, coughing and talking, a lot of droplets of liquid are released into the air, inside which the microorganisms are contained. These droplets can be suspended for hours in the air, i.e. form persistent aerosols. In this airborne way, infection occurs with many acute respiratory diseases (influenza, measles, whooping cough, diphtheria, pulmonary plague, etc.). A similar pathway for the spread of pathogens is one of the main reasons for the development of not only an epidemic, but also large pandemics of influenza, and of pulmonary plague in the past.

In addition to the airborne droplets, the pathogens can spread through the air with the "dust". This is explained by the fact that the microorganisms located in the discharge of patients (drops of sputum, mucus, etc.) are surrounded by a protein substrate, therefore, they are more resistant to drying and other factors. When such drops dry, they turn into a kind of bacterial dust containing many pathogenic bacteria. The bacterial dust plays a particularly important role in the epidemiology of tuberculosis, diphtheria, tularemia and other diseases.

2.4. Food microflora

The content of proteins, carbohydrates, vitamins and other nutrients in food products, favors not only the preservation, but also the reproduction of various microorganisms. In lactic acid and fermented foods there are a large number of microbes that give them taste and a certain consistency. In addition, the products may contain microorganisms or their spores from the environment.

The multiplication of certain microorganisms leads to unsuitability of food for consumption. 25% of the world's products do not reach the consumer due to their spoilage by microbes in most cases. In some cases, food products can be seeded with salmonella, shigella, staphylococci, botulism clostridia, E. coli, B. cereus, Cl. Perfringens and other bacteria that lead to foodborne infections and other diseases in humans.

46

Milk can be infected with the bovine type of mycobacteria, salmonella, brucella, pathogenic streptococci and staphylococci from sick animals.

Meat infection occurs in vivo as a result of illness of animals and birds, as well as during slaughter, butchering, improper storage and transportation of carcasses. Usually, Cl.Perfringens, B.cereus, enterobacteria, fecal streptococci, etc. are found in meat. Most often, meat and meat products, especially minced meat, become infected during their processing, when the microbes come from the surface of the meat grinder, hands and kitchen items.

Fish is infected with a wide variety of microbes that come from water, scales, intestines, as well as from the hands of people processing fish products, and various items (knives, tables, boards, etc.). The most dangerous microorganisms are botulism clostridia, which produce exotoxin in canned foods.

Vegetables and fruit are seeded with shigella, salmonella, vibrio cholerae and microflora, located in the soil and on the hands of people, packaging, transportation and sale in commercial establishments. The improperly canned vegetables, mushrooms, etc. are sometimes the cause of botulism

Various, including pathogenic, microflora (salmonella, fungi, and staphylococci) penetrate the egg powder quite often.

Bread products are relatively rare for the propagation of pathogenic microorganisms. Only the products prepared from wintering grain in the field cause alimentary toxic aleukia caused by pathogenic fungi from the genus Fusarium.

Of all the agents that cause food poisoning in humans, 79% are pathogenic bacteria. Salmonella, staphylococci, streptococci are of particular danger. When multiplying and accumulating in food products, they do not lead to a change in their organoleptic properties.

2.5. Microflora of the body of a healthy person

Human microflora is the result of the mutual adaptation of micro-and macroorganisms in the process of evolution. Most of the bacteria of the constant microflora of the human body adapted to life in its certain parts.

47

With the development of virology and the improvement of virological technology, our ideas about the microflora of the human body have expanded. It was established that not only open cavities, but also tissues of the human body are populated with viruses that are released into the external environment with milk, saliva, sputum, sweat, urine, and feces.

2.6. Skin microflora

Sarcins, molds, and yeasts, as well as some pathogenic and conditionally pathogenic bacteria live on the surface of the skin. Their nutrition is provided by secretions of fat and sebaceous glands, dead cells and decay products. On the surface of the man skin were found from 85,000,000 to 1,000,000,000 microorganisms. The most commonly exposed parts of the human body are infected, mainly the hands.

Violation of the sanitary-hygienic regime, normal working and living conditions of people is often the cause of pustular and fungal skin lesions.

2.7. Mouth microflora

Saliva is an important nutrient substrate for germs. It contains amino acids, proteins, lipids, carbohydrates, inorganic substances.

A few hours after the birth of a baby, microbes can multiply in the oral cavity, and several days later, certain types of streptococci, actinomycetes, lactobacilli, etc. are found.

The most permanent inhabitant of the oral cavity is Str.Salivarius. An especially great number of it are on the surface of the tongue. Most types of oral microorganisms are aerobes and optional anaerobes.

2.8. Gastrointestinal tract microflora

With normal functioning of the stomach, the microflora is almost absent in it, since the bactericidal properties of gastric juice are extremely pronounced. However, the degree of acidity of gastric juice is not always constant. It varies depending on the nature of the food and the amount of water consumed by a person. From the mouth, lactic acid bacteria Sarcina ventriculus, Bac.subtilis, yeast, etc. enter the stomach along with food. In some cases, the penetration of dysenteric, typhoid,

48

paratyphoid bacteria, cholera vibrios and other pathogenic microbes into the stomach, and then into the intestines is possible.

Relatively few microbes are present in the duodenum and small intestine. A huge number of microorganisms are contained in the large intestine. An adult person emits about 17 trillion organisms along with excrement daily.

The intestinal tract of newborns in the first hours of life is sterile. Subsequently, a specific bacterial flora consisting of lactic acid bacteria (bifidobacteria, Lactobacillus acidophilus) is established in the intestines of newborns. It has antagonistic properties to many microbes that can cause intestinal disorders in infants.

In the intestinal microflora of adults, more than 260 species were found. The microbes antagonists (acidophilic, Bulgarian bacteria, actinomycetes, etc.) bring great benefits to the body: they inhibit the development of pathogenic bacteria, which, together with infected food, air and water, can enter the intestines.

2.9. Respiratory tract microflora

Together with air, a man inhales a huge amount of dust particles and microorganisms adsorbed on them. It was found that the number of microbes in the inhaled air is 200-500 times more than in the exhaled one. The alveoli of the lungs and terminal branches of the bronchi are usually sterile. With the weakening of the body's defenses as a result of cooling, exhaustion, vitamin deficiency, injuries, the permanent residents of the respiratory tract become able to cause acute catarrh of the respiratory tract, tonsillitis, pneumonia, bronchitis, etc. The mucous membrane of the nose produces mucin and lysozyme, which have a bactericidal effect. However, despite this, the nasal cavity has a relatively constant microflora. In addition to bacterial microflora, many viruses, in particular adenoviruses, can persist in the respiratory tract for a long period without causing pathological processes.

The microflora of the human body is unstable, it varies in its species composition depending on the age, nutrition and condition of the microorganism.

Violation of the species composition of normal microflora under the influence of infectious and somatic diseases, as well as the result of prolonged and irrational use, which is characterized by the change

49

in the ratio of various types of bacteria, a sharp decrease in the number of bifidobacteria, an increase in staphylococci, antibiotics leads to a state of dysbiosis.

Currently, a new branch of biology is developing, i.e. gnotobiology, which studies the microbial life of macroorganisms. The gnotobiotes, in comparison with ordinary animals, have the enlarged cecum, the underdeveloped lymphoid tissue, they have less mass of internal organs, blood volume, reduced water content in tissues and antibodies in blood serum.

2.10. Animal food microbiology

- Milk microbiology

Milk is the secret of mammary glands of the mammals. Milk contains fatty acids, amino acids, minerals, vitamins, milk sugar and a large number of enzymes. The nipple canal communicates with the external environment from which the microorganisms can enter, and milk serves as a good breeding ground for some microorganisms.Most microbes occur in the nipple canal, the milk tank. Some microbes die under the influence of bactericidal substances contained in milk; only more persistent micrococci and streptococci are preserved, which are close to lactic streptococci in their properties. Microbes, accumulating at the nipple canal, form a plug which can contain the pathogens of infectious diseases. Therefore, the first portions of milk must be drained into a separate bowl to prevent contamination of all milk.

A large number of microbes are found in cow's milk, which has mastitis, in which staphylococci, streptococci, Escherichia coli and other microbes are found. The dirtier the skin of an animal is, the more germs there are that pass into milk. Microbes enter the skin surface from feed, bedding, manure, air. In 1 ml of cow's milk with dirty skin there can be from 170 thousand to 2 million microbial cells, in a cow with clean skin - 20 thousand cells, in those with the systematic cleaning the number of microbes is reduced to 3 thousand. Microbes can get into milk and from the air, from the animals sick with tuberculosis, salmonella, etc.

A great role in the seeding of milk is played by flies. The surface of the body of flies contains a huge number of cells of microorganisms, among which there can be found the pathogenic ones.

50

Phases of change in the microflora of milk during storage.

The composition and number of microorganisms in milk varies during storage. There are several phases.

The antimicrobial phase is characteristic of freshly milked milk, in which all the microorganisms that enter it do not develop. The antimicrobial substances of milk have only a static effect: they inhibit the development of microbes and do not destroy their cells.

The antimicrobial properties of milk are associated with its content of lysozymes, lactenins, bacteriolysins, antitoxins, agglutinins and other substances that come from the blood or are synthesized by the mammary gland.

The activity of the antimicrobial substances depends on the purity of the product and its storage temperature: with the increasing temperature, the activity decreases, and at 55°C the inactivation occurs.

An increase in the number of microorganism cells by a few thousand in 1 ml at the same storage temperature reduces the phase duration by about half. The dependence of the duration of the antimicrobial phase on temperature is below.

Temperature...............0 5 10 25 30 37

Phase duration, h....48 36 24 6 3 2

The phase of mixed microflora lasts 12 - 18hrs. During storage, the antimicrobial substances of milk are gradually destroyed, and the higher the storage temperature is, the faster they do it. With the end of the antimicrobial phase, all the microorganisms entering the milk begin to develop in milk.

Depending on the storage temperature of milk, three types of microflora are distinguished in this phase: cryoflora, or low-temperature flora - 0 - 8°C; mesoflora, or flora of medium temperatures - 10 - 35°C; thermoflora of high temperatures - 40 - 45°C. If at the beginning of the mixed microflora phase the milk is pasteurized, then the psychotropic microflora will be destroyed and the quality of milk will be preserved.

Mesoflora consists of mesophilic microorganisms, the development of which consists of mesophilic microorganisms, the development of which occurs in milk stored at a temperature of 10 - 35 °C.

51

The phase of lactic acid bacteria is observed only at a storage temperature of milk above 10°C. The moment of a marked increase in acidity and the predominance of lactic acid bacteria are conventionally taken as the beginning of the phase: more than 50% of the total number of bacterial cells. The full manifestation of this phase is the absolute predominance of lactic acid bacteria, an increase in acidity up to 60° T and more, and fermentation of milk. All other groups of bacteria stop their development and gradually die.

As a result of the lactic acid process, self-purification of milk occurs, the number of lactic acid bacteria approaches 100%. These are lactic acid products, not fresh milk.

The phase of mold and yeast is conclusive. In this phase, due to the products of their vital activity, lactic acid bacteria cannot withstand low pH and die by the end of the phase.

This creates the conditions for the development of mold and

yeast. Mycelial and non-mycelial fungi develop: milk mold, penicillas, yeast, etc. The fungi use lactic acid, decompose proteins to form alkaline products, as a result of which the pH rises and bacteria become oily.

Butter microbiology

Butter is the most important product of milk processing. It contains the residual microflora of pasteurized cream, as well as the microflora that got into the oil from the outside during its manufacturing. Basically, the bacteria are represented by spore species, sporeless rods and micrococci, among which there are those that produce enzymes that break down milk fat and proteins. The seediness of the surface layer of the oil block is higher than its inner layer.

At the above-zero temperature (15°C) of butter storage, the number of microorganisms increases in it. At a low above-aero temperature (5°C), mainly microorganisms with anti-lytic enzymes grow.

Types of butter spoilage. The types of buter spoilage include rancidity, bitter taste, unclean smell, and mold. Butter rancidity is due to the development of microorganisms that secrete the lipase enzyme. These are, as a rule, mold fungi Oidium lactis, Cladosporium botori, etc. The butter acquires a bright yellow color. Fat hydrolysis products are oxidized to form peroxide compounds, butyric acid is subsequently formed,

52

aldehydes and ketones, which render the butter a characteristic taste and smell of a rancid myrrh oil.

Unclean, dung, and other odors are caused by bacteria of the Escherichia coli group.

Cheese microbiology

Cheese is a very valuable product of milk processing in taste and nutritional properties. The quality of cheese is mainly determined by raw materials - milk, and most importantly – by its purity.

Microorganisms can enter the cheese from the external sources only in a short period of its production - before formation.

The bitter taste is the result of the accumulation of peptides in cheese during the development of micrococci. Cheese bloating results from the release of excess gases (CO_2 and H_2). The causative agent is the bacteria of the Escherichia coli group.

The causative agents of late bloating are cheesy bacteria, which begin to develop in the cheese when the lactic acid process stops and the pH is exceeded.

Meat microbiology

Microorganisms, as a rule, are not found in the blood, muscles and internal organs of healthy animals. This is confirmed by the results of microbiological studies of the slaughter products of healthy animals, killed and opened in compliance with the rules of sterility.

If these rules are violated, then during the slaughter of animals they receive meat and internal organs containing a different number of saprophytic microorganisms, including putrefactive bacteria, coliform bacteria, mold spores, yeast, streptomycetes, cocci, and in some cases salmonella and other pathogenic microorganisms. Seeding of organs and tissues of animals with microorganisms occurs endogenously and exogenously.

Endogenous seeding can occur both during life and after slaughter. The intravital seeding of meat occurs in animals with infectious diseases. Most often this happens when animals are tired during transportation or transfer to meat processing plants. This is due to the fact that the intestinal wall becomes permeable to microorganisms contained in the gastrointestinal tract. Therefore, animals are allowed to rest for at least 3 days before slaughter. During this time

53

glycogen content increases in the muscles, which leads to the increase in the amount of lactic acid and the resistance of meat to putrefactive microbes. The amount of glycogen is one of the factors contributing to the preservation of meat.

Exogenous seeding occurs during the slaughter of animals and subsequent carcass cutting operations.

With a low level of sanitary condition in the slaughter and further carcassing works, the number of microorganisms can reach hundreds or even millions per 1 cm2 of carcass surface area, against not more than several thousand per 1 cm2 subject to sanitary and hygienic rules.

Staphylococci and micrococci, coliform bacteria, various types of putrefactive aerobic bacilli, anaerobic clostridia, yeast, lactic acid bacilli, spores of streptomycetes and molds are most often found.

With the active multiplication of microorganisms as a result of their vital functions, soreness, rotting, acid fermentation, pigmentation, mold and glow can occur. The main causative agents of mucus are aerobic bacteria of the genera Pseudomonas and Achromobacter. At a temperature of 5°C during the storage period, micrococci, streptococci, Streptomyces Uybkjcnyst multiply. When meat is stored under aerobic conditions, the psychrophilic bacteria of the genera Lactobacterium, Microbacterium, and atromonas cause mucus.

The rate, at which the sagging occurs, depends on humidity and storage temperature.

When storing meat with signs of mucus, its further damage occurs, called rotting, which is caused by non-spore-forming aerobic bacteria Bac.Prodigiosum, Pr.Vulgaris, Ps.Fluorescens, as well as spore-forming aerobic Bac.Subtilis, Bac.Vesentericus, Bac.Mycoides and anaerobic bacteria .Sporogenes, Cl. Putrificus, Cl. Perfringens.

Further the putrefactive bacteria penetrate the thickness of the meat and cause the breakdown of muscle tissue.

The acidic fermentation is accompanied by the appearance of an unpleasant odor, gray or greenish-gray color in the section and softening of muscle tissue. Psychrophilic lactobacilli of the genus Lactobacterium, bacteria of the genus Microbacterium, and yeast, which can develop in meat, appear. These microorganisms break down carbohydrates of muscle tissue with the release of organic acids.

54

Pigmentation is the appearance of colored spots on the surface of meat as a result of reproduction and the formation of colonies on the surface of meat with various pigments. The causative agents are aerobic or facultative anaerobic microorganisms: Ps. Fluorescens, Ps. Pyocyanea, Ps. Cyncyanea, Bact.Prodigiosum, sarcinol, pigment yeast.

The mildew rarely appears when observing the temperature and humidity conditions of storage. Mold fungi during development on the surface of meat, as a rule, do not cause profound changes in it, but they can create more favorable conditions for the subsequent development of putrefactive bacteria.

The luminescence arises as a result of the multiplication of photogenic bacteria on the surface of the meat, which have the ability to glow - phosphorescence.

A typical representative of photogenic bacteria is Photobacterium Phosphoreum, i.e. a motionless cocciform bacillus. Photogenic bacteria develop well on fish and meat, but do not cause any changes in smell, texture, or other organoleptic characteristics.

Sausages microbiology

During the preparation of sausages, the minced meat is seeded with microorganisms. The main source of seeding is the raw material of sausages, therefore, high sanitary requirements are imposed on it. Only after cutting and deboning, the meat seed rate grows 100 times or more. Of the microorganisms, putrefactive, enterococci, streptomycetes, yeast, mold fungi, E. Coli, Proteus, staphylococci are found.

When salting, the source of seeding by microorganisms can be salt containing salt-tolerant and salt-loving microorganisms: Bac.Mesentericua, pigment cocci, yeast, mold spores, streptomycetes.

The natural intestinal membranes contain a large number of different microorganisms, many of which are the causative agents of spoilage of meat and meat products.

Bac.Halophlum, Micr.Carmens, Mic.Rpseus halophilus, ic.Citreus, Mic.Albus, Subtilis, Bac.Mesentericus, Bac.Mycoides, streptomycetes, etc.are often found in them.

Types of sausages spoilage. With improper storage of sausages, the residual microflora of sausages begins to multiply and cause different types of spoilage.

55

The rotting of sausages due to the activity of putrefactive bacteria: Ps.pyocyarica, Pr.Vulgarus, Bac.Subtilis, Bac.Mesentericus, Cl.Sporogenes. The putrid decomposition of sausages occurs simultaneously throughout their whole length.

Rancidity most often occurs with prolonged storage of smoked sausages. Ranopharynx is the result of reproduction in the sausage of the following microorganisms: Bact.Prodigiosum, Endomyces lactis, Cladosporium herbarum and others with lipolytic enzymes. Lipolytic enzymes break down fat into glycerin and fatty acids, which are oxidized.

The resulting aldehydes and ketones give the product a rancid taste and pungent odor. acid fermentation is caused by Cl.Perfringens, lactic acid bacteria, yeast, etc. This type of spoilage is more common in cooked and liverwurst sausages, which contain a lot of carbohydrates (flour, vegetable impurities) and have high humidity. Mold is due to the storage of uncooked sausages in high humidity.

Poultry meat microbiology

Poultry meat is also a favorable environment for the development of microorganisms. In birds, especially waterfowl, salmonella, pathogens of toxic infections, can be found in muscles.

Half-gutted carcasses of birds are usually significantly seeded with microorganisms than the gutted ones.

As a result of storage of carcasses at the temperature of 1°C, the first sign of damage may appear, that is the odor. The microflora of the carcass surface at this time consists mainly of aerobic, non-spore sticks of genera Pseudomonas (up to 70 - 75%), Acinetobacter, Moraxella. There are facultative anaerobic bacteria Aeromonas, Enterobacter, and E. coli.

Eggs microbiology

Seeding of eggs with microorganisms can occur endogenously and exogenously. In endogenous seeding, microorganisms penetrate the egg during its formation in the ovary or oviduct of a sick bird.

56

Of particular danger are the eggs of waterfowl, which are infected by S.Enteritidis, S.Cholera suis, S.Typhi-murium, S.dubllin,

A.Anantrum, and others. In this regard, it is forbidden to sell the duck and goose eggs through the stores, markets and catering. The exogenous insemination of eggs is associated with the contamination of the shell with droppings, soil, litter, feathers, etc.

A layer of mucus is deposited on the surface of the shell when the eggs are laid, which, when dried, forms an overshell film - a cuticle which includes lysozyme, which has bactericidal properties against many microorganisms. If the cuticle is damaged, microorganisms penetrate the eggs through the pores in the shell. Bacteria of the genus Pseudononas and Proteus penetrate from the surface of the shell into the egg on the 2nd - 5th day, Salm.Typhmurim - the 8th - 11th day, e.coli - on the 13th - 15th day, Aspergillus – on the 5th - 9th day.

Waterfowl eggs are often a source of tuberculosis and salmonella infection. The greatest danger among salmonella is represented by bacteria of the species S.Typhimurium, which infected not only the duck, but also the chicken eggs. In addition to salmonella, cholera vibrio and other pathogenic microorganisms, including tuberculosis pathogens, can enter the egg through the pores of the shell.

Egg products microbiology

Eggs are used to make egg melange and egg powder. The melange is made either in the form of a mixture of proteins and yolks in natural proportion, or in the form of a yolk mass freed from shell and protein, or in the form of an egg beck.

The melange refers to perishable products. A large number of microorganisms are found in it. The composition of this microflora depends on the purity of the eggs. The main source of seeding of melange by microorganisms is the shell. Eggs must be disinfected.

Most often, various types of cocci, mold fungi, Pr.Vulgaris, Bac.Subtilis, Bac.Mesentericus, E.coli, and salmonella are found in the finished melange.

The thawed melange should be used within several hours, otherwise it will deteriorate.

Egg powder is obtained by drying the egg mass by spraying it in special kA and frozen egg melange at a temperature not exceeding 60°. Chicken table eggs are used for this.

57

Egg powder should contain (%) not more than 9 moisture and 4 water; not less than 45 protein substances and 35 fat. In a viable state, spore-forming bacteria, staphylococci, streptococci, ... are found in a viable state. During storage, partial death of microorganisms is detected. The shelf life of egg powder at a temperature not exceeding 20 ° C and a relative humidity of not higher than 75% is about 6 months

The advantage of egg powder is reduced volume and weight, the ability to store in an uncooled room, good transportability.

Fish microbiology

The surface of freshly caught sea fish contains the most bacteria of the family Achromobacteriacea, which make up 60% of the total microflora.

The microflora of freshwater fish in the central part of Russia primarily consists of psychrophilic microorganisms of the genera Pseudomonas, Aeromonas, Achromobacter, Micrococcus. In addition, staphylococci can get on fish during processing, as they make up 40% of the microflora of the hands and nasopharynx.

Frozen fish is of great importance, as it allows you to continuously provide the population with fish. In frozen fish e.coli, Staphylococci, Salmonella, the causative agent of botulism are found. To obtain frozen fish that is safe from the point of view of sanitation, fresh fish processed in strict compliance with sanitary and hygienic requirements should be used for freezing.

Salted fish is in great demand, therefore, its sanitary and hygienic condition is of great importance. There are three types of salting: soft, medium and strong. With soft salting in the muscle tissue of fish, the content of salt should not exceed 10%. With an average salting salt is 10 - 12%. With strong salting, the percentage of salt in muscle tissue is 14%.

Salted fish contains mesophilic microorganisms that can reproduce even at a temperature of 5C. The cause of the appearance of brown spots on the surface are molds of the genus Sporendonema. The growth of molds is suppressed at a storage temperature of 5°C.

58

Smoked fish has been used by humans for a long time. There are two types of smoking: hot and cold.

With hot smoking, the temperature inside the fish should rise to 65 ° C within 30 minutes. Almost after the treatment with smoke, the fish meat becomes sterile also because the smoke contains a number of substances with bactericidal properties. However, smoke chemicals do not penetrate the fish meat.

Cold smoking is carried out by smoke at 18-26°C for 2-3 days. In this case, water is removed and the constituent parts of the smoke penetrate the fish meat.

The types of spoilage of smoked fish are wet rot, dry rot and mold. The cause of wet rot is psychrophilic bacteria, while the fish becomes wet, sticky, emits an acute putrefactive odor.

Dry rot is caused by micrococci and aerobic spore-forming bacteria, which remained viable during smoking, yeast and sarcinol. The fish gets a matte shade, muscle tissue becomes loose.

Mold is most commonly found on the surface of fish. The causative agents are mold fungi that enter the fish, both during and after smoking.

Smoked fish poisoning can occur due to the content of salmonella in it, most oftenS.typhimurium and Co.Bptulinum.

Canned fish. Fish is canned by sterilization.

The basis for choosing a sterilization regime is the destruction of heat-resistant spores of Cl.Botulium. The signs of spoilage of canned food are bombing - swelling of the upper and lower lids of cans. The bombing is caused by the gases formed during the decomposition of fish by bacteria Cl.Sporogenes, Roseum, Bac.Cereus, and Bac.Coagulans. Canned fish poisoning is also caused by bacteria Cl.Botulinum.

2.11. Fresh fruits and vegetables microbiology

The microorganisms developing on fruits and vegetables can be divided into three groups.

The first group includes the microorganisms that develop on the fruits, tubers and other organs of plants exclusively during storage

59

and do not damage the plants, but the vegetation season. These are the typical saprophytes. They can cause disease only in the weakened plants through damaged integuments.

The entire development cycle of these microorganisms can take place in storage. The unfavorable storage conditions are too high temperature and humidity, contributing to infection. The microorganisms of the first group include the following:
Rhizopus nigricas - the causative agent of black mold rot of many fruits;
Aspergillus niger - the causative agent of black mold rot of citrus;
Penicillium digitatum - the causative agent of olive mold rot of citrus;
Erwina carjtovora - the causative agent of wet bacterial rot of vegetables;

The second group includes microorganisms that infect plants in the late stages of vegetation in the field under adverse weather conditions. The microorganisms of this group in their development are more closely associated with plants. This group includes the following microorganisms:

Fusarium - the causative agent of potato fusarium;
Phytophtora infestans – the causative agent of late blight of potato;
Sclerotinia libertiana - the causative agent of white rot of many fruits and vegetables, especially carrots;
Botrytis cinerea - the widespread causative agent of gray rot of many fruits and vegetables;
Rhizoctonia - pathogen of root rot.

The third group includes the microorganisms that affect only the vegetative plants. These microorganisms have widely expressed the parasitic properties and are able to infect strong plants. As a result, we can draw the following conclusions:
1. The microorganisms of the second and third groups are able to penetrate into plant tissues through the intact integuments, in contrast to the microorganisms of the first group, which do not have this ability.
2.The microorganisms of the second group can penetrate the plant through the intact integumentary tissues.
It should be noted that in different countries, depending on the environmental conditions, the microflora of fruits and vegetables varies significantly. So, in Italy, England, Germany, the main causative agent of apple diseases in

60

during storage is a fungus Gloesporium album; in the USA – Penicillium expanum; in a number of areas of the Pacific coast of America – Neofabreo malcorticis.
The species composition of microflora during storage of fruits may vary depending on the duration of storage.

2.12. Microbial physiology
The deeper we get into the secret of microorganisms, the more convinced we become that we need to know not only the

morphological features, but also the changes that they make in nature, and this will allow us to target them and regulate physiological processes in the environment.

The structure of the microorganisms cells is constantly changing. The water content in the cytoplasm of most species ranges from 75% to 85%. In the spores of bacilli and clostridia, the water concentration is 40-50%. Water is the main component of the cell by its amount. It is in a free state or connected with other components. The bound water is a structural element of the cytoplasm and cannot be a solvent.

Free water serves as a dispersion medium for colloids and a solvent for crystalline substances, a source of hydrogen and hydroxyl ions, and a participant in chemical reactions.
It serves as a medium in which the movement of ions and electric charges takes place.

Minerals. The composition of bacteria includes inorganic substances (phosphorus, sulfur, sodium, magnesium, potassium, calcium, iron, chlorine, etc.) and trace elements (molybdenum, cobalt, boron, manganese, zinc, copper, etc.). The total content of minerals in bacteria obtained on ordinary nutrient media is in the range of 2-14% of their dry weight.

Proteins. The protein distributed in the cytoplasm, nucleoid, cytoplasmic membrane and other cellular structures makes up 50-80% of the dry matter of the bacterial cell. The proteins contain nucleoproteins. The lipoproteins are inside the cell in the form of inclusions. On the surface of the cytoplasm, the lipoproteins form a membrane that regulates the flow of substances into the bacterial cell. Proteins include enzymes (enzymes), which play a great role in the life of microorganisms.

Nucleic acids. The content of nucleic acids in a bacterial cell depends on the type of bacteria, culture medium and

61

it fluctuates in the range of 10-30% of dry matter. Three types of ribonucleic acid are known: ribosomal acid (r RNA), transport (t RNA), matrix (m RNA). The ribosomal RNA is a part of the ribosomes, the transport one transfers amino acids to the ribosomes, the matrix one provides the sequence for the inclusion of amino acids in the molecule of the polypeptide chain.

DNA consists of adenine, guanine, cytosine, thymine, phosphoric acid, deoxyribase; RNA contains adenine, guanine, cytosine, uracil, phosphoric acid, and ribose. The difference between these two nucleic acids is that the DNA has a nitrogenous base, thymine and deoxyribose, while the RNA contains uracil and ribase.

The quantitative and qualitative diversity of protein substances, their complexes and amino acids endows microorganisms with the species peculiarity.

Carbohydrates. Mono- and disaccharides make up 12-18% of the dry weight. The bulk of carbohydrates is a polysaccharide complex, in a free or bound state with proteins and lipids, contained in the cell membranes and the mucous layer. A number of bacteria in the cytoplasm have a relatively large number of inclusions resembling glycogen or starch in their chemical composition.

Lipids. In bacteria that do not contain fat in the form of inclusions, lipids make up about 10% of the dry residue. In bacteria that deposit fat in the form of special inclusions, the amount of lipids reaches 40%. Bacterial lipids are composed of free fatty acids (26-28%), neutral fats, waxes and phospholipids. Lipids are part of the cytoplasmic membrane and its derivatives. In the cytoplasm they are deposited in reserve.

2.13. Chemical composition of viruses

Viruses contain one of the types of nucleic acids – either DNA or RNA. The ratio of nucleic substances, with other components of the virion, ranges from 1% to 40%. The DNA of many viruses has a ring structure rather than a linear one. The nucleic acids of viruses and cells differ in the composition of their nitrogenous bases and the sugar component of the DNA. Virus proteins are consist of the usual 16 - 20 L - amino acids. In some viruses, protein consists of one type of polypeptide chain, in others it is formed by several types of polypeptides.

62

Some viruses have revealed the presence of the enzymes with the help of which they are reproduced in the cells of plants and animals.

The nature of viruses. The question of the origin of viruses and their nature is the subject of numerous studies and theoretical discussions. Some scientists consider viruses as descendants of ancient non-cellular forms of living parasitic systems, functionally closely related to the host cell, but developing independently and genetically independently of them. Others believe that viruses arose from unicellular organisms, which, as a result of regressive evolution, lost their protein synthesizing systems and became strict intracellular parasites. The rest of the researchers claim that viruses came from cellular elements that became autonomous systems.

One of the central issues of methodological importance is the one of whether viruses can be considered living systems.
Despite the differences of viruses' structure, chemical composition, genetic apparatus from prokaryotes and eukaryotes, they, as well as other living systems, possess all the treats of life, that is, the ability of evolution, self-

replication, mutation, genetical information transfer, specificity of interaction with host cells.

Before entry into a cell, the viruses behave like molecules of gigantic size, and inside cells they become living beings able to reproduce and pass their peculiarities.
Viruses exist in nature in two forms: 1)ecto, i.e.virion and 2)endo, i.e.vegetative (reproductive form).

Phages are viruses of bacteria and a number of the microorganisms, which, under certain conditions cause lysis (decomposition) of their hosts. In 1898, N.F.Gamalea showed that the filtrate of anthrax bacillus causes the lysis of fresh culture of these microorganisms. In 1917, a French microbiologist F. d'Hérelle studied a dysentery agent, watched the lysis of bacterial culture while adding the filtrate of ill people's feces to it.

The lytic principle remained during high passaging of dysentery bacillus culture and it became even more active. The author called the agent dissolving bacteria a bacteriophage, and the bacteriophage's action ending with bacteria lysis was called the bacteriophage phenomenon. F. d'Hérelle made an assumption that the bacteriophage is an infectious agent which lyses bacteria, as a result of which the daughter phage particles reach the environment.

63

On solid media seeded with a mixture of phage with a bacterial culture, sterile spots or negative phage colonies appear in the places of bacterial lysis. The nomenclature of bacteriophages is based on the species name of the host. For example, the phages that lyse dysenteric bacteria are called dysentery bacteriophages, those that lyse salmonella - salmonella bacteriophages, the diphtheria bacteria are called the diphtheria bacteriophages.

The phages are organisms that, like all the living beings, are able to multiply, transmit their properties to offspring and change under the influence of various factors. They can infect and destroy only young developing cells, being their parasites.

Phages are more resistant to physical and chemical factors than many human viruses. Most of them are inactivated at the temperatures above 65-70°C. They tolerate freezing well and are stored for a long time at low temperatures and drying.

The phages have strict specificity, i.e., the ability to parasitize only in a certain form of the microorganism. Phages are usually named after the host microbe.

The interaction of the phage with a sensitive cell passes through successive stages (1-2 hours). After the penetration of the phage NA into the bacterial cell, the latter begins the reproduction of the NA phage protein, and the cell lysis, as well as the emergence of mature phage particles occur.

The virulent phages cause lysis of the infected cell with the release of a large number of phage particles into the environment that can infect new cells.

The moderate phages do not lyse all the cells in a population. Some of them enter symbiosis, integrate into the chromosome of the cell and get the name prophage. The formation of a single chromosome occirs. The bacterial cell does not die. The Prophage, which has become a part of the cell's genome, during its propagation can be transmitted to an unlimited number of offspring, i.e. new cells.

The phages are ubiquitous. They are found in water, soil, sewage, human and animal secretions. Almost all the known bacteria are the hosts of phages specific to them.
The resistance of phages to physical and chemical factors is

higher than that of the vegetative forms of their hosts. The phages withstand heat up to 75°C,

64

long drying, pH from 2.0 to 8.5. They are not sensitive to antibiotics, thymol, chloroform and a number of other substances that destroy the accompanying microflora.

The use of phages is based on strict specificity and the ability to destroy microbial cells or enter into symbiosis with them.

The prevention and treatment of infections with the help of phages is based on the fact that when a patient encounters a pathogen, the phage destroys it. Bacteriophage preparations are used to treat dysentery, salmonellosis, and purulent infections caused by bacteria.

2.14. Microbial antagonism

The antagonistic relationship of microbes has attracted the attention of bacteriologists already at the dawn of the development of this science. Soon after the discoveries of Pasteur, which showed the microbial nature of a number of diseases of man, animals and plants, it was found that other organisms, later called antagonists, can suppress and even kill pathogens.

Although the microorganisms are found on a wide variety of substrates, ranging from atmospheric dust, plant debris, various food substances and ending with living plants and animals, their natural habitat is soil and water reservoirs.

In natural substrates, like soils and ponds, the microbes grow and metabolize in a different way from that in pure cultures. In pure cultures, the microbes are free from the associative or competitive influences of other microorganisms. In the mixed populations of natural substrates, complex ecological relationships arise between microbes.

With the co-existence of two or more organisms, some of them can become antagonistic to others. The composition of the medium and growth conditions affect the activity of antagonists.

A characteristic feature of antibiotic substances is their selective action. The antagonist microbes with their antibiotic metabolites suppress a specific set of microbes. So, for example, staphylococci in the urine can become antagonistic to E. coli, or E. coli can have an antagonistic effect on staphylococci, depending on the initial ratio of the numbers of these two groups, on the formation of metabolic products

65

substances and other factors. The producer of streptomycin Act.Streptomycini suppresses 20 types of bacteria and some types of yeast and fungi. Act.Fluorescens affects neither bacteria, nor mycobacteria, but actively suppresses yeast and some yeast-like and mycelial fungi.

The Bayle's statement is of great interest. He noted that for each type of bacteria there is a certain maximum number of cells capable of existing in a given volume of the medium. Once this concentration is reached, the multiplication ceases regardless of the depletion of nutrients or the formation of toxic metabolic products.

Garré (1887) first noted that antagonism can be one-sided or two-sided. In the first case, one organism suppresses another, which is not an antagonist in relation to the first. In the second case, both organisms mutually suppress each other.

Numerous data show that already at the dawn of the development of bacteriology, it was found that microbes change the environment in which they grow, and make it unsuitable for the growth of other organisms. I.I.Mechnikov emphasized the fact that the bacillus of Massol acts antagonistically on other microbes not only by the formation of lactic acid, but also as a result of the release of special substances by it.

It is necessary to note that the antimicrobial features and the selectiveness of the antagonists actions are a natural or specific property. What is more, this particular characteristic is strictly constant if the same conditions of the experiment are kept the antimicrobial effect accurately repeats.

A characteristic feature of the antagonists forming antibiotic substances is the specific nature of the interaction. The microbes antagonists do not inhibit the growth of cultures of their own species with their antibiotics. They act on an organism belonging to alien species. This feature is well studied in actinomycetes, which confirms the conclusions of I.I. Mechnikov.

The phenomenon of antagonism was first systematically studied by Babesh in 1885. It was found that the typhoid bacillus cannot be isolated from sterile wastewater, in which the antagonist was present for 7 days before.
Many aerobic spore bacteria with antagonistic properties were isolated from soil, sewage, and manure. They are Bac.Subtilis, Bac.Mesentericus and others. They form an alcohol-soluble substance with bactericidal properties on the peptone media.

66

Duclos was one of the first to isolate and describe the antagonistic spore bacteria that were obtained from cheese and named Tirotrics. Rosenthal isolated optionally thermophilic antagonistic bacteria from the soil and manure, belonging to the group of Bac.Mesentericus, capable of dissolving living and dead bacteria.

It is important to note that the bacterial antagonism is a property of individual strains, and not of individual bacterial species. It was found that the antagonistic properties do not increase with passivation. The blue pus bacillus has an antagonistic effect on the microbes of anthrax. If two different types of microbes are introduced into a liquid nutrient medium, one of which is an antagonist, for example, plague and the blue pus bacillus (the latter will be an antagonist), then after a few days it is no longer possible to detect a plague bacillus that will die, displaced by the blue pus bacillus.

The mechanism of antagonists' action is different. They propagate themselves quicker and, being less demandful, they surpass the growth of the symbiote and, therefore, suffocate it. Further, there was finally confirmed that the blue pus bacillus is also active in relation to Escherichia coli, Mycobactérium tuberculósis and other bacteria. A preparation, called pyocyanase, possessing strong bacteriologic action was obtained. The preparation could bear the heating by flowing steam during two hours. The pyocyanase in an animal body turns into a high molecular weight protein which preserves its bacteriologic property.

It should be remembered that when a microbe is continuously exposed to the antibiotic substance of its cohabiting antagonist, it develops immunity or resistance, which is fixed hereditarily and transmitted from generation to generation. This way, the insensitive species of bacteria are formed. Penicillin becomes harmless to the bacteria adapted in it, since they have acquired the ability to synthesize the antidote - penicillinase and other metabolites.

It is noted that sulfamide compounds are inactivated by bacteria. It can be assumed that microorganisms have many other substances that are produced in some cases constantly, in others temporarily, in the presence of an antagonist-stimulus in the substrate. It turned out that microbes can change the permeability of the membranes, trap the substance in the membrane or in separate areas of the protoplasm (cytoplasm).

67

In September 1989 in India, 500 cases of typhoid fever have been identified. In 83% of cases, the bacteria that caused the typhoid fever were resistant to chloramphenicol, a special medicine that was the main treatment for typhoid fever in India. Twelve child deaths were noted. Consequently, the resistance to the preparation arose.

Act.streptomycini, the preparation of auroomycin has the same antimicrobial spectrum that is characteristic of culture Act.Aureofaciens. The same can be said about other antibiotics: penicillin, terramycin, neomycin, grisin, subtilin and others. The previously reported data on the antimicrobial action of antagonists almost exactly repeat the data obtained in experiments with chemically accurate preparations.

Thus, the resistance of microorganisms to antibiotics and, in general, to preparation has become an accute problem in medicine and veterinary medicine. The antagonistic relationships are established among viruses when one virus protects the body from the introduction of another virus into it. In virology, this phenomenon is called the interference of viruses.

The criterion for the sensitivity of microorganisms to antibiotics is the minimum concentration of the antibiotic, which inhibits (delays) the growth of the pathogen under standard experimental conditions.

Actinomycetes as antagonists. The actinomycetes are found everywhere in nature: in soil, at the bottom of rivers and lakes, in dust particles and on the surface of plants. The actinomycetes form mycelium. But more often these are unicellular organisms. Of the microbes of antagonists, most attention has recently been paid to the actinomycetes. According to many researchers, the actinomycetes dissolve cells of not only living, but also dead cultures. That is why a group of ray fungi, as antagonists and producers of antibiotics, is considered the most interesting and promising one. The extremely widespread of the actinomycetes in nature suggests that these organisms play an important role in the circulation of substances, both organic and mineral. Among the actinomycetes there are aerobes and anaerobes, mesophiles and thermophiles. They are mainly saprophytes.

Actinomycetes suppress various microorganisms: bacteria, fungi, yeast, algae, protozoan organisms, insects, etc. There are also such cultures of actinomycetes that suppress phages and viruses with their antibiotics. Rosenthal isolated actinomycete from the air, which was a true idiologic antagonist of diphtheria bacillus.

68

The most famous are the actinomycetes, which produce antibiotics: streptomycin, aureomycin, terramycin, neomycin, chloromycetin, sarcomycin, and actinomycin.

Fungi from the penicillium group form the penicillin, which has found a wide practical application.
Penicillium notatum has a strong bactericidal, bacteriostatic effect on staphylococci, streptococci, diphtheria bacillus, gonococci, and meningococci. The first antibiotic, penicillin, was discovered by the English microbiologist A Fleming in 1929. The discovery of penicillin was a great victory of modern biological, medical and chemical sciences in the fight against various infectious and inflammatory processes, as well as a powerful incentive for the search for new antibiotics synthesized by various groups of microorganisms.

Interestingly enough, various fungi act antagonistically on the representatives of their own species, or on the fungi of other species. The same relationship is observed between viruses. If two strains of influenza viruses are simultaneously introduced into the tissue culture, then one of them, which is introduced in larger quantities, inhibits the development of the other.

Antagonists are widespread in the nature. They are found everywhere where life exists. Most of all they are found in the soil. The microbial population density of soils varies with the changes in soil fertility and environmental conditions.

The phenomenon of antagonism served as the basis for the emergence of the science of antibiotics – the chemicals synthesized by microorganisms which inhibit the growth of other microorganisms or cause their death.

The antibiotics or antibiotic substances were proposed by Waxman in 1942. Currently, the antibiotics should be called the chemical compounds formed by various microorganisms during their life, as well as the derivatives of these compounds, which are capable of selectively inhibiting the growth of microorganisms in small concentrations or causing their death.

The action of antibiotics is based on their suppression of the synthesis of proteins and nucleic acids. Such antibiotics as streptomycin and neomycin inhibit the binding of amino acids to each other. Erythromycin disrupts the function of the ribosome subunits. Tetracycline inhibits the attachment of aminoacyl t-RNA to ribosomes.

69

Mitomycin C selectively inhibits the DNA synthesis, without initially affecting the synthesis of the RNA and proteins.
Actinomycin D forms a complex with the DNA and disrupts the synthesis of the RNA polymerase and inhibits the synthesis of the bacterial t-RNA. Penicillin disrupts the formation of the cell wall in bacteria.

According to some evidence in the literature, some antibiotics inhibit the function of microbial cell proliferation. The latter continue to develop and grow, stretch in length, increase in size. Their division does not occur or occurs late. Various deformed cellular elements are obtained: spherical, bulbous, filiform, amoeba-shaped, etc., often reaching enormous sizes. Such giant cells often have a clearly degenerative or inflationary character of the internal structure.

Gigantism of cells is observed when exposed to various antibiotic substances: penicillin, streptomycin, aureomycin, subtilin. gramicidin, etc.

The process of the cell size reduction to the extent of appearing the filterable forms can take place under the influence of the antibiotic substances.

The results of numerous researchers have shown that antibiotic substances are adsorbed by the bacterial cells and cause one or another reaction in them. The absorption of penicillin by the cells of bacteria has been studied by many researchers. It has been established that to inhibit growth, the cell must absorb a certain amount of antibiotic.

The phenomenon of adsorption should be considered as the first stage of the interaction between the cells and antibiotics. The ability of cells to absorb and concentrate antibiotics is not only a physical, but also a chemical or, rather, a biochemical process.

Some researchers suggest that antibiotics, when entering the bacterial cells, bind to certain components and form a toxic substance. The latter incapacitates one or another function leading the cells to death or suppression.
The effect of antibiotics on the cellular enzymes of the system has been proven by numerous studies. The antibiotic substances either delay the formation of enzymes or inhibit their activity.
Antibiotics are indispensable therapeutic agents and are used to treat a large number of infectious diseases of both people and

animals.

70

Besides, in agriculture, certain antibiotics are used as the animal growth promoters.

Antibiotics are also used in the food and cereal industry. Good results are obtained when the fish is immersed into the sea water chilled to 1 – 1,5 °C with only 2 ml of chlortetracycline per liter in it. You can extend the shelf life of fresh fish by 5 days or more when storing fish on ice, which contains 1 - 2 mg of chlortetracycline or 25 mg of biomycin per 1 liter.

With the widespread use of antibiotics as therapeutic agents, a rapid accumulation of resistant forms of microorganisms to these compounds occurs, and this requires the replacement of one antibiotic with another, i.e. search for more and more new antibiotics. As a result, the health authorities are very careful about the use of antibiotics, since with repeated intake with food even negligible amounts the persistent forms of pathogens can appear in the human body due to the displacement of beneficial microorganisms from the normal microflora of the gastrointestinal tract by antibiotics.

In the imbalance of the intestinal microflora, there are 4 stages of dysbiosis: 1) a moderate decrease in the number of obligate bacteria in the intestinal cavity; 2) a critical decrease in the number of bifidobacteria and lactobacilli, which leads to the rapid growth of pathogenic bacteria; 3) inflammatory lesions of the intestinal walls, persistent diarrhea, undigested food; 4) there is a general depletion of the body, anemia, vitamin deficiency.

At a medical forum in Novokuznetsk, it was said that with dysbiosis in the intestine there is not only a shortage of saprophytes bacteria, but also the most important thing, a pathologically large number of fungi. Therefore, in the treatment of dysbacteriosis, you can never do without antifungal drugs that must be taken with bifidobacteria.

For the preservation of food products it is allowed to use special antibiotics that are not used in medicine. Such an antibiotic is, for example, nisin, synthesized by lactic streptococci, which inhibits the growth of staphylococci, some streptococci, anaerobic thermophilic spore bacteria, causative agents of canning damage.

71

Nisin, which inhibits spore germination, is used in the manufacture of condensed milk and processed cheeses.

2.15.Phytoncides

For food industry specialists, the antimicrobial properties of plants are of great importance - phytoncides secreted by some plants, as well as tissue juices that cause the death of ciliates, bauteria, yeast, and molds. Phytoncides were found in the representatives of all the groups of higher plants. The greatest antibiotic properties arepossessed by the phytoncides of onion, garlic and some other plants. The biological activity of phytoncides varies in the same plant species depending on the season of the year. The phytoncides in plants are either distributed diffusely across all tissues, or concentrated in certain areas.

Loshchenko and Lomakin established the presence in the chicken egg of an antimicrobial substance - lysozyme, which causes a bacterial cell crisis. It was found in saliva, tears, the secretions of the nasal mucosa, and the blood serum.
Lysozyme is found in milk, the liver, the spleen of warm-blooded animals. To date, a large number of different phytoncides have

been studied. Some of them are obtained in the chemically pure form. Alicin is a human phytoncide. In the intact garlic retains antibiotic activity for a year or more. It has been shown that in garlic, allicin is contained not in the form of a free substance, but in the form of a compound that can be converted into an antibiotic.

Alicin inhibits the development of gram-positive and gram-negative microbes and the development of tubercle bacillus.

2.16. Microorganisms genetics

The doctrine of the heredity and variability of organisms was founded by Ch. Darwin in 1859, which proved that all the species of plants and animals that exist on the Earth appeared through alterability.

The basic laws of genetics are discovered and formulated by the Czech natural scientist G. Mendel. A certain contribution to the doctrine of the variability of microorganisms was made by L. Pasteur, L. S. Tsenkovsky, I. I. Mechnikov and others. Mechnikov approached the study of alterability from a Darwinian perspective.

72

In particular, he wrote, "It was in the field of microbiology that the possibility of changing the nature of bacteria by altering the external conditions was proved. Moreover, lasting inherited alterations can be achieved."

Many researchers adhered to this concept, but there were monomorphists (R. Koch). They claimed the constancy of microbial species. They denied alterability under the influence of the external environment.

The genetics of microorganisms is the basis of molecular biology. The most important problems of molecular genetics are studied on microorganisms. The molecular biology owes its achievements to the success of the genetics of bacteria and viruses. Many discoveries in the field of genetics of microorganisms have been very effectively used in biology and medicine.

Further studies showed that under the influence of the medium the organisms often acquire new features which can be temporary or permanent, inherited. The microbes lose virulence, acquire drug resistance, change morphological and cultural properties.

Temperature, chemicals, antibiotics, and other environmental factors can cause morphological changes.

The sticks take a rounded shape, become longer and thicker, form a swelling. Morphological changes are more often observed in old cultures, when the products of vital activity of world organisms accumulate in large quantities. A variety of signs and properties in individuals and groups of individuals of any degree of kinship is the alterability inherent in all living organisms. Therefore, in nature there are no individuals that are identical in all respects and properties. The theory of alterability is also used to denote the ability of living organisms to respond morphophysiologically to external influences.

Bacteria and viruses have become the main objects of study of the structure and function of genes only because they are relatively simple to organize, reproduce fairly quickly, are usually haploid, and induced mutations and genetic recombinations can be easily reproduced at the bottom. Microbial alterability is not limited to morphology, size and cultural characteristics. It affects other properties. In particular, the alterability of the enzymatic ability of bacteria.

73

N.F. Gamaleya (1859-1949) observed morphological changes in a number of microbes in the form of the formation of giant balls, amoeba-shaped forms, thickened threads, etc.

N.F.Gamaleya

He called this phenomenon heteromorphism. Its essence lies in the adaptation of bacteria to unusual conditions of existence.

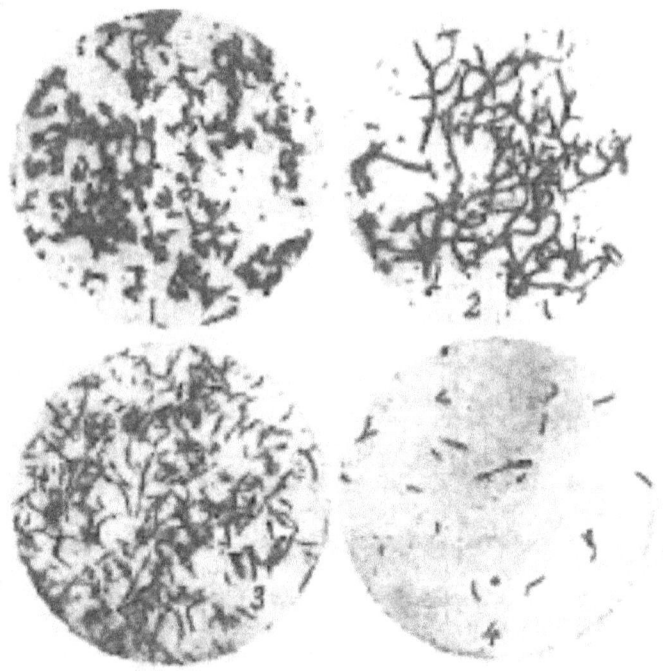

Fig.6. Bacterial morphology alterability

74

It should be noted that any change in morphological characters is accompanied by a change in physiological properties, therefore, the division of the types of bacterial alterability into morphological, cultural, enzymatic, biological is conditional and is done in order to more clearly cover the diverse material on this issue.

Change in cultural properties. Along with morphological abnormalities in microbes, the changes in cultural properties are also often observed. The same microbes under the same conditions can have different cultural characteristics. If the medium is dense, smooth (S-shape), then the colonies are formed to be dull, transparent with smooth edges.

If the medium is rough (R-shape), then the colonies are rough, opaque, with a folded surface.

Moreover, such cultures differ not only in form, but also in other ways. S-form is more pathogenic and is characterized by good agglutinating properties. Such alterations, in which there is a separation, the splitting of characters in microbes, is called dissonance. It is believed that mutations are at its core.

Alterability of biological properties. A very important circumstance is that in pathogenic microbes, under the influence of various factors, the degree of their pathogenicity changes. A decrease in the pathogenicity of microbes while maintaining their ability to induce immunity was noted a long time ago, but L. Pasteur was able to reproduce such properties for the first time. He managed to obtain a live vaccine against anthrax by exposure to elevated temperature on the culture of anthrax bacilli.

In 1885, L. Pasteur changed the properties of the causative agent of rabies through 113 consecutive infections of rabbits in the brain; it turned out to be harmless with subcutaneous administration and at the same time able to prevent the development of the disease in persons bitten by a rabid animal. It is no coincidence that L. Pasteur drew attention to a test tube, with cholera culture (of birds) forgotten in a thermostat, which did not cause disease in chickens, but the immunity was created after its introduction into the body.

Alterability of enzymatic functions. The microbial alterability is not limited to morphology, size, and cultural characteristics; it also affects other properties, in particular, the variability of the enzymatic ability of bacteria.

Adding a certain substance to the medium can cause the activation of an enzyme that was previously in a latent state.

The action of some toxic substances on bacteria can deprive them of the ability to produce various enzymes. The cultivation of Cl.Perfringens in a low iron environment leads to a decrease in the enzymatic ability of this microbe.

Under the influence of the corresponding metabolite (inducer or repressor), the synthesis of a specific enzyme is induced or repressed. These processes are subject to genetic control, due to which the cells regulate their physiological activity according to environmental conditions by the enzyme synthesis activation and disactivation. Adding a certain substance to the medium can cause the activation of the enzyme that was previously in a latent state.

Thus, we are convinced that by the exposure to various factors, the biological properties of microorganisms can be changed. We are currently witnessing an increase in the resistance of microbes to drugs. Some microbes even feed on antibiotics.

The phenotypic changes include adaptation and modification. Adaptation is the adjustment of microorganisms to the environmental conditions. Modification is a change in the microorganism under the influence of environmental conditions. So, for example, when calcium chloride is added to the medium, E. coli cells are greatly shortened. If this substance is removed from the medium, they again take their original form. Therefore, the modification is a reaction to external stimuli that are not associated with a violation of physiological processes in the body.

Genetic or hereditary changes are the result of mutations or recombinations of genes. The mutations are common to all living beings, including microorganisms. They appear as a result of changes in the sequence of the DNA bases, as well as the nucleides in the genes, and are inherited. Such a gene encodes a protein that differs from the original one in properties and functions.

Mutations are the persistent hereditary changes in the properties of a microorganism (morphological, cultural, biochemical, biological, etc.) that are not associated with the recombination process.

76

Mutations can be accompanied by loss (deletion) or addition (duplication) of one base or a small group of bases in the DNA molecule.

The bacterial mutations are divided into: 1) spontaneous, arising under the influence of external factors, without the researcher's intervention; 2) induced by the treatment of the microbial population with mutagenic agents (radiation, temperature, chemical and other influences.

Spontaneous mutations are very rare, about one in a hundred thousand. They are characterized by a change in any one sign and are usually stable. Some microorganisms (.E.coli, S.tiphi, etc.) have genes that are stable to mutagenic effects, but change in vivo, which is probably an important factor in the divergence of enterobacteria.

The induced mutations, or mutagenic mutations, occur due to the exposure to the environmental factors. They are relatively common. The mutagens are divided into: physical, chemical and biological. Different types of radiation belong to the physical ones: ultraviolet, x-ray, and radioactive. They cause damage to the genetic apparatus, a change in the characteristics, properties of microbes. The chemical potent substances include:

poisonous (mustard), medicinal (iodine, hydrogen peroxide), acids (nitrous) and others.

An example of biological mutagens may be the DNA. So, when some types of oncoviruses are introduced into the cells of the drosophila embryo, the adult individuals acquire new features: unusual outgrowths or depressions appear on the head, and sometimes the eyes disappear. The mutant effect of viruses and live vaccines on mammals has been proven. They damage the hereditary apparatus. At the same time, a technique has been developed for obtaining useful traits of microorganisms. Thus, highly active strains of producers of antibiotics and other substances were isolated. In particular, strains were obtained that increased the biosynthesis of penicillin by 10 thousand times, vitamin B2 (riboflavin) by 20 thousands, vitamin B12 by 50 thousand times. The same method exponentially the yield of essential amino acids (lysine, glutamic acid).

Genetic recombinations. This group of variability includes the recombination of genes that occur as a result of transformation,

77

transduction and conjugation. The genetic recombination of bacteria and viruses leads to the appearance of recombinants that have the features of both parents: the main set of recipient genes and a certain part of the donor genes.
The recombination creates an inexhaustible source of variability, since in vivo the microorganisms are in a state of various associations - biocenoses and parasitocenoses, which creates favorable conditions for the genes exchange.

The transformation is the process of transferring a portion of the genetic material of DNA containing one pair of nucleotides from a donor cell to a recipient cell. In this case, 5 stages are passed: 1 – the DNA adsorption on the surface of the microbial cell; 2 – the DNA penetration into the recipient cell; 3 – the pairing of the introduced DNA with the chromosomal structures of the cell; 4 – the inclusion of a portion of the DNA of the donor cell into the chromosome structures of the recipient; 5 - a further change in the nucleotide during subsequent divisions.

The transformation is possible only when the recipient has a special state of "competence" in which he is able to perceive the donor DNA.

It was found with various types of microorganisms (hay bacillus, Haemophilus influnzae, meningococci, etc.) that during transformation, the genes that determine antibiotic resistance, phagolysability, and other properties can be transmitted from the donor to the recipient. The effectiveness of the transformation depends on a number of conditions (medium composition, temperature) of the physiological state of the recipients and the transforming DNA.

The question arises whether the transformation between microorganisms of different species is possible. As it was already noted, the appearance of recombinants is preceded by the recombination of the transforming donor DNA with the recipient cell gene.

During the transformation of two closely linked genetic markers located on the same DNA molecule, the transformants with single features appear more often than the transformants that carry two features simultaneously.

The discovery of the phenomenon of transformation allows a deeper study of the genetic role of the DNA at the molecular level and clarifies a number of issues regarding microbial association in alteration, infection and immunity.

78

Transduction is a change in which the genetic material from a donor cell to a recipient cell carries a transducing (moderate) phage, that is, a phage that does not cause its destruction.

The transduction mechanism is as follows. During the reproduction of some mild phages, small DNA fragments of the donor bacteria are inserted into the phage genome, which transfers them to the recipient bacteria.

There are three types of transduction: general (generalized), specific and abortive. With the general transduction, the transfer of different or several characters can occur simultaneously.

The specific transduction is characterized by the transfer of only a specific trait. With the abortive transduction, the manifestation of a new feature is not observed. It should be noted that the abortive transduction occurs 10 times more often than the transduction, accompanied by the integration of the genetic material of the donor and recipient.

The interspecific transduction is very important. With the help of viruses – the carriers of genes from one living system to another, a gene was transplanted from a nitrogen-fixing bacterium into a bacterium that does not have this property.

The conjugation of bacteria is a form of the sexual process through the direct contact, which resembles a reduced sexual process (Fig. 7). The conjugation process is determined and controlled by special plasmids – the fertility factors. A cell containing at least one of these plasmids acquires the properties of a donor, and deprived one of it – of a recipient.

Plots of the DNA transferred from the donor find homologous regions in the recipient's DNA molecule, between which the genetic exchange takes place. As a result, a part of the donor DNA is inserted into the genome of the recipient, and the corresponding part of the recipient DNA is excluded from it.

In the conjugation of bacteria, there is only a partial exchange of genetic information of conjugating individuals, in contrast to the conjugation of plants.

79

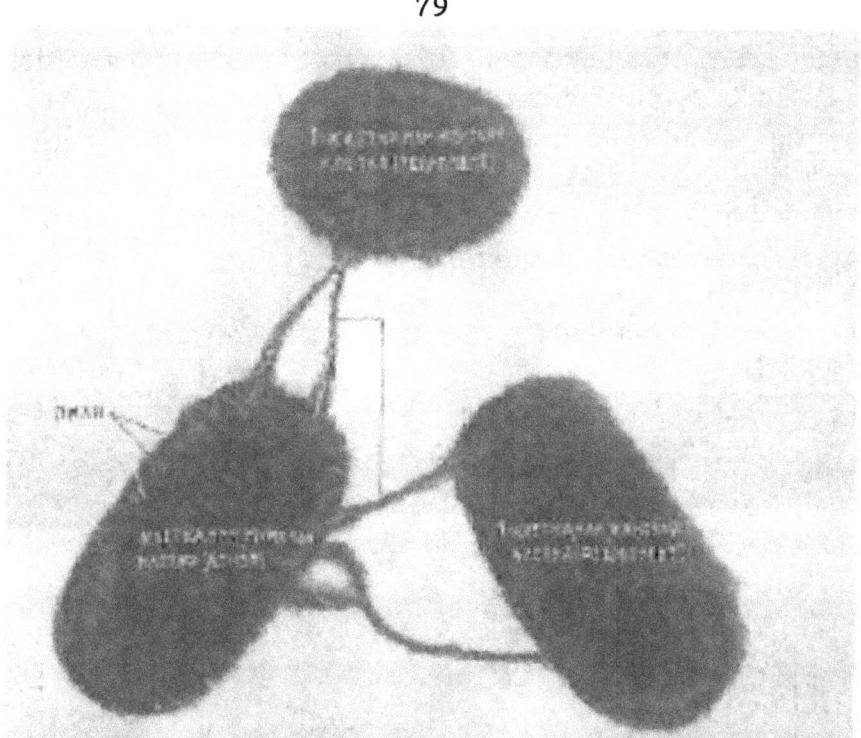

Fig. 7. Conjugation of bacteria
Micrograph of conjugating bacteria of one
"male" and two "female" individuals

During conjugation, a unilateral transfer of genetic material from the donor to the occurs recipient. As a result, forms with features of conjugating cells are formed. It has been established that the transfer of genetic material from a donor cell to a recipient cell occurs in a certain direction and with a known frequency.

The F-pili (long thin protein tubes) were found in gram-negative bacteria. Through the F-pili, the transfer of genetic material occurs; F-pili serve as a conductor of a hereditary substance. During conjugation, the transfer of not the whole nucleotide, but only of its parts occurs. This process takes place in a certain sequence within 30-90 minutes.

Currently, a new direction in molecular biology - genetic engineering - has developed. It is engaged in the construction, isolation and transplantation of certain genes from one cell to another. As a result, the cells acquire new properties that can further be used in the national economy.

A gene synthesizing insulin was extracted from the human body and transferred to the genome of E. coli.

80

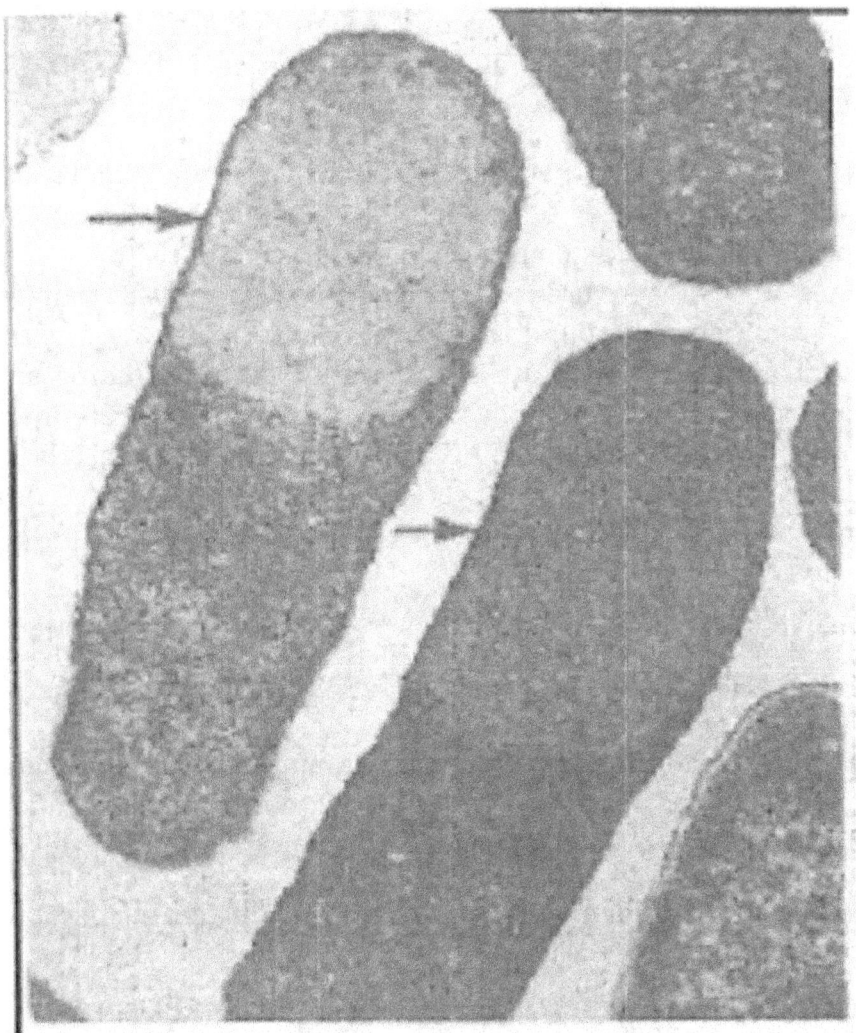

Fig. 8 Biosynthesis of human insulin in the
genetically engineered E. coli cells.

Such a bacterium is capable of producing a protein hormone -
insulin. Thus, you can get enough insulin, which lacks so much.

2.17 Genetically Modified Food Sources

Transferring the genes of any organism into the recipient's
cell allows to obtain plants, animals, or microorganisms with

recombinant genes and, accordingly, new properties. In 1994, a genetically modified, storage-resistant tomato appeared in the US food markets.

Subsequently, corn, potatoes, soybeans, pumpkin, and sugar beets were received for use in the United States, Japan, and the European Union countries. In Russia, modified soybean of lines 40 - 3-2 were used. Currently, hundreds of GMS (sources) of food are used around the world. Over the past 2 years, the cultivation area of cultured transgenic plants, including soy, rape, tomatoes, potatoes, has increased more than by 20 times, and this trend is progressing. It should be noted that plants of transgenic modification become resistant to herbicides, insecticides, viruses, acquire new consumer advantages.

81

At the same time, the amount of applied herbicides and insecticides decreases, their residual content in the products decreases, which leads to the increase in product quality.

In Russia, more than 150 items of GMP (products) are produced. Soya is used in the production of more than 3,000 food products: soups, baby cereals, potato chips, margarines, salad dressings, canned fish and much more. In Australia, grapes are obtained from which wine is produced with improved organoleptic properties. New technologies for producing transgenic agricultural animals and birds, aimed at increasing their productivity, are gaining importance.

The Russian Academy of Agricultural Sciences has developed the technologies for obtaining a number of genetically modified agricultural crops. The key point is to study the chemical composition of new food products, while studying the biochemical indicator of blood, urine, morphological studies of organs, the immune status of the body. These studies are conducted on animals. If necessary, special studies are conducted:

- the study of allergic properties;

-the identification of the possibility of mutagenic and carcinogenic effects;

-the assessment of the possibility of individual studies, including embryotoxic. The final stage is the testing of new products on volunteers.

In all countries, the registration of GMP has one goal, that is, to reliably assess the safety and usefulness of new analogues of traditional products. An undesirable effect of GMP is the possibility of transformation of the transferred genetic material. On July 1, 1999, a special procedure was introduced for biomedical assessment and registration of food products obtained from the GMP, which provides for mandatory state registration of food products and food raw materials.

In Russia, the law "On State Regulation in the Field of Genetic Engineering" No. 86 dated 07/05/96 was adopted. According to the law, one of the main tasks of state regulation is to determine the mechanism that ensures the safety of citizens and the environment in the process of carrying out the genetic engineering activities and using its results.

82

2.18 Morphology of fungi

Fungi are one of the largest and most prosperous organ groups. It includes about 100,000 species. Fungi differ from plants and animals primarily in the type of food. They do not have chloroplasts and cannot synthesize organic substances themselves, but can only utilize the organic substances of dead organisms. Fungi are eukaryotes that have lost chlorophyll.

The vegetative body of the fungus is represented by mycelium or spawn consisting of highly branched filaments - hyphae. Fungal hyphae are unicellular with a large number of nuclei representing one giant cell, and multicellular, or septic, i.e. divided by septa into separate cells containing from one to many nuclei.

Fungi absorb nutrients throughout the surface by diffusion. This distinguishes them from animals that first swallow food and then digest it. Thus, the digestion of fungi is external, carried out by extracellular enzymes. The spore formation allows saprophytes to easily spread to other products, such as Penicillium. Mushrooms parasites can be optional or obligate, more often parasitizing on plants than on animals. The obligate parasites, as a rule, do not cause the death of their hosts, while the facultative ones do this often, and then live saprophytically on dead remains.

The fungi statistics is periodically improved. Currently, they include 5 classes: zygomycetes, ascomycetes, basidomycetes, deuteromycetes - imperfect fungi   Chitridiomycetes.

Actinomycetes (actis - ray, micis - fungus), or radiant fungi. This is a group of unicellular gram-positive microorganisms. They have a branchy shape. Among actinomycetes, there are saprophytes that take part in soil-forming processes, but there are others that cause the infectious diseases.

Actinomycetes are one branched cell whose hyphae form mycelium (mycelium). Mycelium can be substrate and air. At the ends of the mycelium the spores (conidia) form. It is they which provide the reproduction.

83

In actinomycetes, we can find the features of fungi and bacteria. In one case, we have unicellular mycelium (an organism). They are similar in size to bacteria (examined under an immersion microscope system). They are stained with aniline dyes and positive Gram, the ability to grow on meat and peptone agar at a temperature of 35-37C. They have a prokaryotic type of cell. As parasites, actinomycetes cause severe diseases in humans and animals (actinomycosis). Currently, the name radiant fungi is considered obsolete. On a nutrient medium they form a colony of substrate mycelium and aerial mycelium in the form of fluffy velvet or powdery deposits. Sporiferous hyphae (spore-carriers) form on the filaments of the aerial mycelium.

The actinomycetes unite about 700 species. They are distributed in soil, ponds, in the air and on nutrient residues. They can be found in the oral cavity, plaque, tonsil lacunae.

The **zygomycetes** unite more than 1000 species. This is a subclass of lower fungi. They live in the soil on the decaying remains of plants and animals. These are unicellular organisms with highly developed mycelium. They multiply sexually and asexually. The zygomycetes unite more than 1000 species. This is a subclass of lower mushrooms. They live in the soil on the decaying remains of plants and animals. These are unicellular organisms with highly developed mycelium. They multiply sexually and asexually. They form vertically growing fungi - sporangios. They have negative geotropism. The sporangien bearer turns into sporangia. The protoplasm of sporangia is divided into two parts, then around each of such parts its own cell wall appears and a spore containing several nuclei is formed.

When sporangia is ruptured, the spores enter the external environment and, falling into favorable conditions, give rise to a new mold. This is the ordinary bread mold. It also grows on apples and other fruits.

The sexual stage of reproduction in lower fungi begins with the formation of germ cells or gametes, which are formed in differentiated cells - gametangia. The gamete fusion can occur both in gametangia and outside of it. Gametangia do not differ in size. This process of sexual reproduction is called isogamy. After nuclear fusion, a zygospore with a lot of diploid nuclei is formed.

Ascomycetes - one of the classes of higher fungi has 25-35 thousand species. Ascomicota - marsupials - the most numerous and a relatively highly organized group of fungi.

84

These include yeast, a number of common molds, real powdery fungi, morels and truffles.

The genus of Aspergillus is included into the class of ascomycetes. This also includes the Asp.Niger species, which is widespread in nature. It lives on wet objects, bread, jam, etc. A. Fumigatus, A. Flawus, and A. Niger cause diseases in humans. Some species produce aflotoxin, which has a carcinogenic effect.

More than 40 species are described in patients with various clinical forms of aspergillosis. In humans, they cause damage to the skin of the trunk, limbs, adnexal nasal cavities, lungs, bronchi, corneas of the eye, external auditory meatus, sometimes bones and other organs and tissues.

The genus of Penicillium, which is widespread in nature, belongs to the class of ascomycetes. It is found in feed, dairy products, on wet objects. Some species (Penicillium chrysogenum and others) are used to make penicillin. More than 30 species are pathogenic to humans. They cause damage to the skin, nails, ear, upper respiratory tract and lungs, as well as a generalized infection with the formation of foci of internal organs. The asexual reproduction of penicillium is carried out using conidia, Fig. 6. These are the spores that form at the end of special hyphae called conidiophores. They are not enclosed in sporangia, they are freely dispersed during ripening.

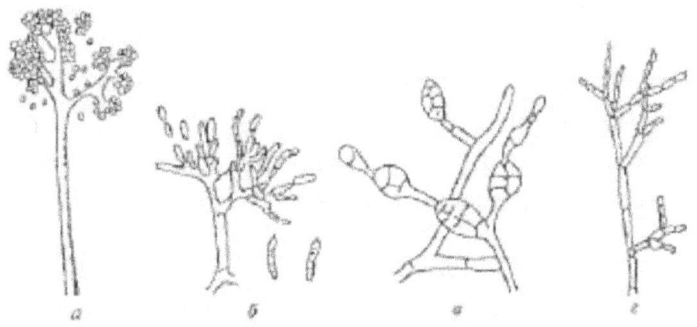

Fig. 9 Conidiophores

a – Botrytis; b- Fusarium; c- Alternaria; d- Cladosporium
Basidomycetes - Basidomyicota - are higher fungi (edible). The number of species amount to about 100 thousand. These are saprophytes and parasites of plants, animals and humans.

85

Oriental medicine (Vietnam, China, Japan) used higher fungi (basidomycetes). Thus, the Lin Chi fungus is an object of medical research in oncology. These higher fungi are the most highly organized. Their large fruiting bodies immediately attract attention, whether they are edible fungi or toadstools. These also include numerous obligate parasites - rust and smut fungi. It should be noted that the mycelium of pileated fungi grows saprophytically on the organic material of the soil and can live there for many years.

Chitridiomycetes. The fungi of this class have the poorly developed mycelium. The vegetative body is a single cell, sometimes devoid of a wall. The sexual process is represented by the gametolamy of various types.

Many chitridiomycetes are intracellular plant parasites. The most important representative of this class is the synchitrium fungus (Synchytrium endobioticum) - the causative agent of potato tuber cancer. The cancer-infested potatoes cannot be sold. Many chitridiomycetes parasitize on planktonic algae and higher aquatic plants.

Deuteromycetes or imperfect fungi (Deuteromycetes). This class includes the fungi with well-developed mycelium that do not have sexual reproduction. This is one of the largest classes of mushrooms, including about 30% of all known species.

Deuteromycetes are widespread in nature in all the areas of the globe. Many of them live in soil, are found in abundance on plants, debris, less often on substrates of animal origin. They take an active part in the decomposition of organic residues and in the soil formation process. Some of them cause moldy foods and spoilage of various industrial products. Some deuteromycetes cause diseases in animals and humans (germatophytes).

Developing on grain and other food products, the individual deuteromycetes, for example, the representatives of the Fusasrium genus emit toxins in them, which can cause severe poisoning when using such products by people or animals.

The most common causative agents of food spoilage are such deuteromycetes as botritis, fusarium, alternaria, cladosporium, oidium, and monilia. Botritis affects apples, pears, vegetables, and especially berries.

86

The surface of the fruit and vegetables is covered with a fluffy gray coating, the tissue turns brown, becomes watery, and softens.

The diseases of fruits, vegetables, and grains caused by Fusarium are called the Fusarium infections. Some species of Fusarium form mycotoxins.

2.19. The value of fungi in the food industry

According to the data of technical processes, connected with the food spoilage caused by fungi, it should be mentioned that fungi play an important role in a number of food industry branches.

The founder of microbiology, the outstanding L. Pasteur, making experiments with yeast more than a century ago, discovered the microbiological synthesis of protein from ammonia and organic compounds that do not contain nitrogen.

A modern person consumes 800g of food and 2000g of water per day. The daily diet of the population of our planet is more than 4 million tons of food. Meanwhile, it is estimated that the rate of agricultural products production will further lag. Now the shortage of food in the world exceeds 60 million tons. A particularly acute issue is the lack of protein, vitamins and minerals.

It has been scientifically proved that it is impossible to eliminate a huge shortage in food products due to the expansion of cultivation areas and increase of the livestock quantity. Therefore, the scientific basis of the modern food production strategy is the search for new resources, the search for new sources of protein. Topical issues are the selection of the most productive species of fish, marine animals, and other seafood.

Great success has been achieved in the production of vitamin preparations and their premixes. Especially promising is the biotechnology industry - the use of microorganisms as the sources of individual components of food products. It is microorganisms that will solve the problem of protein and vitamin deficiency. This is a high reproduction rate of microorganisms. In the world of living beings, microorganisms have no match in the rate of protein and vitamins production.

The terms for the protein mass doubling: cattle - 5 years, pigs - 4 months, chickens - 1 month, higher plants - more than 3 months, bacteria and yeast - 1-6 hours. A promising direction in solving this problem is genetic engineering. These are the genetically modified sources.

87

Yeast is widely used in the production of bakery products. In the manufacture of dough from wheat flour, the extruded baking yeast is usually used. It is produced at specialized yeast plants.

Beer - a low alcohol drink - is obtained by fermentation with special races of yeast called Saccharomycts cerevisiae. The same race of yeast is used to produce the best varieties of wines and champagne.

Fermented-bread kvass - our national drink, is a product of unfinished alcoholic fermentation, where this yeast is also used. Yeast is an indispensable inhabitant of milk and dairy products. In fresh milk, after a few hours, the percentage of yeast in the total number of microorganisms is 13%.

Yeast is a necessary component of the fermentation starter cultures used to obtain kefir, koumiss and other national drinks. Alcoholic fermentation underlies the production of acidophilus-yeast milk and various whey drinks. For this purpose, both lactobacillus yeast species and those incapable of using this sugar are used.

The biomass of fungi in the soil is about 90% of the mass

of all microorganisms and invertebrates. The calorie content of dried mushrooms is 220 kcal per 100 grams of their air-dry mass. Recently, in the food industry, the methods of preserving mushrooms by freezing and freeze-drying are widely used.

2.20. The structure and propagation of yeast

In the classification of fungi, yeast occupies a special place, they are the unicellular fungi, which belong to the class of Ascomycetes or sac fungi. They are called so because their spores form in sacs. Yeast is the most extensive class of fungi, which includes about 35,000 species.

Yeast cells are immobile, sometimes forming the so-called false mycelium. Yeast is widely distributed in nature (in soil, on the surface of plants, fruits, berries and vegetables, in the variety of substrates containing sugar) and has been used by humans since ancient times in the food industry, winemaking, brewing, baking and other industries related to fermentation.

88

The most widespread method of vegetative propagation of yeast is budding (Fig. 10).

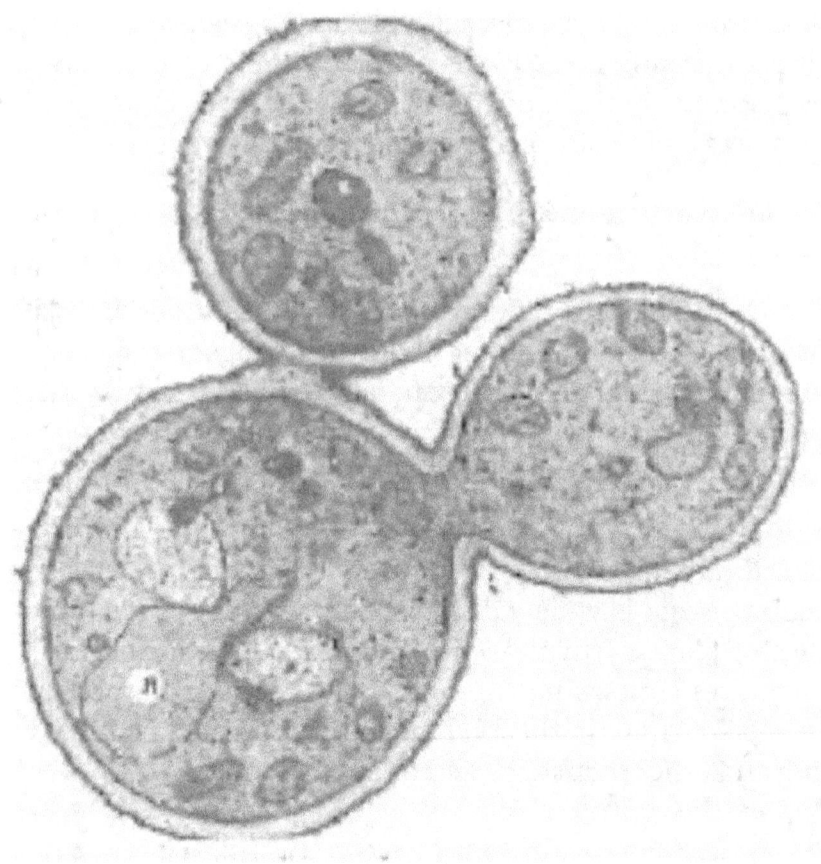

Fig. 10: Electronic photograph of a budding cell of
brewer's yeast (Saccharomyces cerevisiae)

The first daughter cell has not yet separated from the original
cell, on which the next bud is formed. In the parent cell, as well
as in the newly formed buds, mitochondria are visible.

In the formation of the yeast biomass, the complex enzymatic
reactions occur. They ensure the formation of protein and
vitamins from carbohydrates contained in a nutrient medium.

The yeast cells have the most diverse shapes: round, oval,
oval-ovoid, elliptical, cylindrical, lemon-shaped, sickle-shaped,

bulbous (Fig. 11).

Yeast is divided into three families based on the vegetative propagation: the Saccharomycetacea family, the Schizosaccharomycetacea family, and the Saccharomycodacea family.

The Saccharomyces cerevisiae species includes the yeast used in bakery, alcohol production, brewing, winemaking, and production of kvass.

89

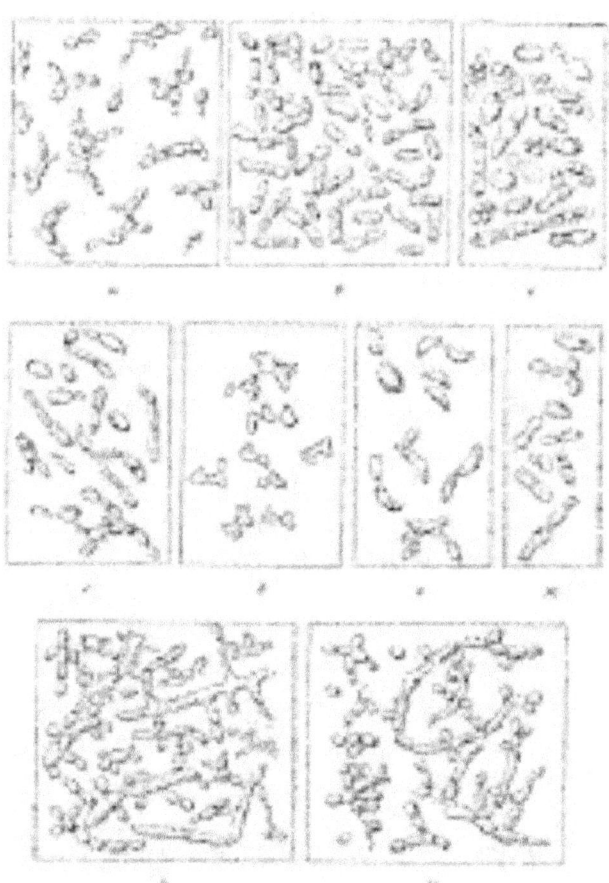

Fig. 11. Forms of yeast cells.

a - oval ovoid, b - cylindrical, c - apicular, lemon-shaped,
d - arrow-shaped, e - triangular, f - sickle-shaped, g
- bulbous, h, i - elongated, mycelium-shaped

Yeast propagates both vegetatively and sexually. The most common method of vegetative propagation is budding. Some types of yeast reproduce by binary division. It should be noted that during the development on liquid media, the cell walls of some yeasts may become mucinous and stick together with each other, forming flakes settling to the bottom. Such yeast is called flocculent. The yeast, the cell walls of which do not become mucinous, is called powdery. The powdery yeast is in suspension during the entire fermentation process. The flocculent yeast cells are larger and heavier than those of the powdery ones. They give a smaller increase in biomass, but have a higher fermentation activity and ferment the wort more fully, form higher alcohols. The flocculent yeast creates a better aroma of the drink.

The strains of Saccharomyces cerevisiae are subdivided into the races of bottom and top fermentation. The majority of wine and brewer's yeast belongs to the bottom fermentation race, while the top fermentation races include alcohol, baker's, and some brewers. The bottom fermentation yeast functions in the production at the temperature of 6-10°C and lower, and the top fermentation yeast - usually at 14-25°C.

90

At the end of fermentation, the bottom yeast works to the bottom, forming a dense precipitate, the top yeast comes to the surface and form a "cap".

The length of the yeast cell does not exceed 10 - 15 μm. The yeast is eukaryotic. The structure of the yeast cell is similar to the structure of the fungal cell. Its cell wall has a layered structure. The number of layers is usually three, but sometimes it reaches 10 or more. The wall is a strong outer cover of the yeast cell.

The cytoplasmic membrane is located behind the cell wall. It consists of lipoproteins and protects the cytoplasm from external influences. It regulates the penetration of nutrients into the cell, and provides homeostasis.

In the process of ATP oxidation or hydrolysis, the bacteria transport H ions to the surface of the cytoplasmic membrane and, thereby, bring it into the so-called energized state. The resulting gradient of the electrochemical potential of H ions is a driving force for many processes: active transport into the bacteria of most of the compounds entering it (including cations and anions), the movement of bacteria, the performance of certain types of chemical work. Ion exchange provides stabilization of pH values inside bacteria.

It should be noted that the transport of ions into the microbial cells is only one of the few sections of the modern membrane science that reflects the mechanisms of penetration of charged particles through biological membranes. This process has many similarities with the entry of organic compounds into the bacterial cell.

The cytoplasm is a semi-liquid colloidal phase that has all the cell structures. It is the cytoplasm that provides a direct connection between the organelles of the cell.

The endoplasmic reticulum is located in the cytoplasm. This is a whole system of channels, vesicles or cisterns associated with the cytoplasmic membrane and nuclear wall. It provides transport of substances throughout the cell. It synthesizes antibodies. Ribosomes are located in the cytoplasm of the cell. They contain 50% of all ribonucleic acid. In the ribosomes, the process of protein synthesis occurs.

The Golgi apparatus is either a group of bubbles of various diameters, or parallel discoid plates. This system is connected to the network of the endoplasmic reticulum. They perform an excretory function.

91

It is responsible for the synthesis of membranes. The Golgi apparatus membranes are sites for the lysosome formation that protect the cell from damage by the decay products by foreign agents.

Mitochondria provide the cell with energy. The energy is transferred to the adenositriphosphate system. The mitochondria are located in the cytoplasm. The synthesis process takes place in them.

The nucleus is isolated and surrounded by a nucleoleme. The direct connection of the nucleoplasm with the cytoplasm is through the nuclear pores. The proribosomes migrate from the nucleus to the cytoplasm. The main component of the nucleus is chromatin. Here the RNA cells are synthesized (Fig. 12).

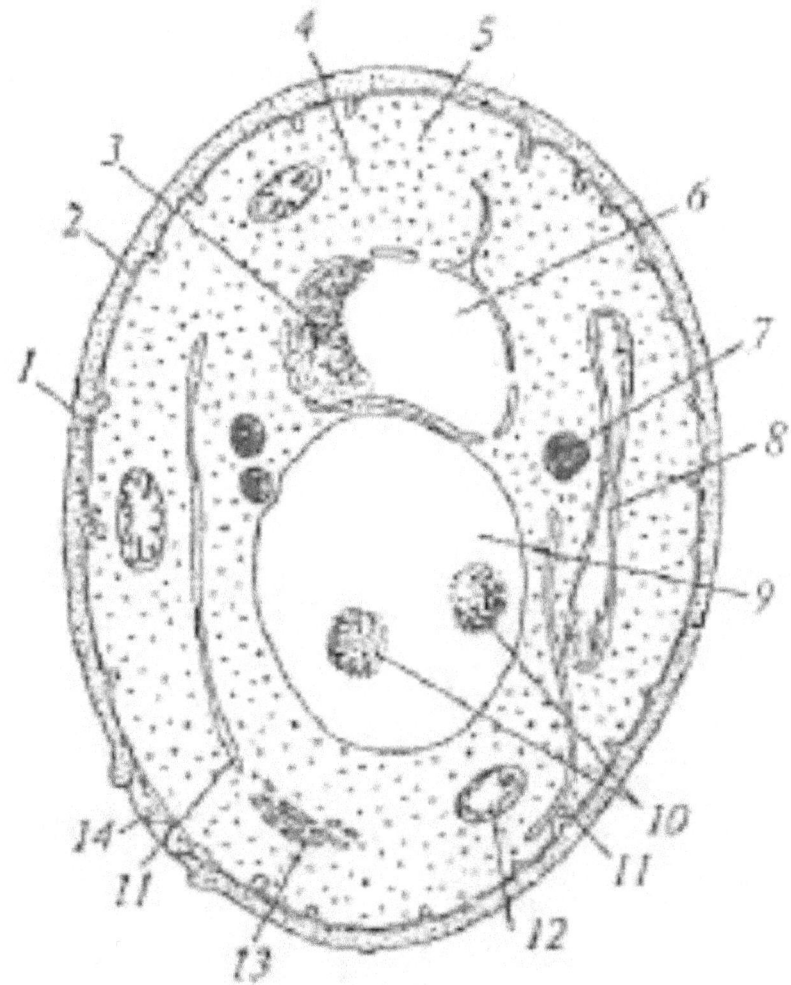

Fig. 12. Schematic representation of a
cross section of a yeast cell.

1 - cytoplasmic membrane; 2 - cell wall; 3 - nucleolus; 4 -
cytoplasm; 5 - ribosomes; 6 - core; 7 - fat droplets; 8 - the Golgi
apparatus; 9 - vacuole; 10 - granules of polyphosphate; 11 -
endoplasmic reticulum; 12 - mitochondria; 13 - dictosome; 14 –
scar, remaining at the place where the daughter cell has budded.

The Saccharomyces cerevisiae strains are subdivided into races of the bottom and top fermentation. The majority of wine and brewer's yeast belongs to the bottom fermentation race, while the top fermentation races include alcohol, baker's, and some brewers. The bottom fermentation yeast functions in production at the temperature of 6-10°C and lower, and the top ones - usually at 14-25°C. At the end of the fermentation, the bottom yeast works to the bottom, forming a dense precipitate, the top yeast floats to the surface and forms a "cap".
The cells of Saccharomyces cerevisiae have an elliptical shape (Fig. 13).

92

Fig. 13 Yeast of the Saccharomyces cerevisiae genus

According to the behavior in the fermentation medium, the yeast is also divided into flocculent and powdery. The basis of this division is the difference in cell agglutination. The flocculent ones at the end of fermentation stick together into lumps and either work to the bottom, or rise to the surface.

The powdery yeast is in suspension during the entire fermentation process. The flocculent yeast cells are larger and heavier than those of the powdery ones. They give a smaller increase in biomass, but have a higher fermentation activity and more fully ferment the wort, form higher alcohols. The flocculent yeast creates a better aroma of the drink.

2.21. Baker's yeast production.

The baker's pressed and dry yeast is produced at the specialized yeast plants. The nutrient medium for their cultivation is the brightened beet molasses. The cultivation temperature is maintained at about 30 °C, the pH of the medium is 4.5 - 5.5. When the maximum number of cells per volume of the medium is formed, they are drained, separated from the medium, washed with water, concentrated, and pressed to a moisture content of 73-75%. At present, the dry yeast with the moisture content of 8 - 10% is also produced. The disadvantage of the pressed yeast is its short shelf life,

93

and the disadvantage of the dry yeast is the loss of a significant part of the activity during the drying process. Now the yeast of the third generation is produces. It can improve the quality of bread and actively resist the process of staling.

2.22. Yeast food industry pests

The beer wort is a complete breeding medium for most microorganisms. But beer has natural resistance against microorganisms. Hop resins have a bactericidal effect, low pH (5.4 - 4.6), low temperature and anaerobiosis. In addition, ethanol and carbon dioxide inhibit the development of microorganisms. However, there are microorganisms that can develop in brewing conditions and be dangerous, as they reduce the quality of beer. These include acid-resistant, alcohol-bearing, anaerobic organisms. Their breeding in beer causes a change in smell, taste, the formation of turbidity, increased acidity. They come from air, water, raw materials, crops, equipment, containers and from personnel.

The main raw material for the production of beer is malt, the quality of which depends on barley. Both bacteria and mycelial fungi are found on the surface and under the shell. There are tens and hundreds of thousands of cells in one gram of harvested grain. The filamentous fungi in this case are 1-2%. They are represented by the genera of Altenaria, Fuzarinum, Helmintosporium, Cladosporium. The grain affected by fungi significantly loses its qualities. Their livelihoods adversely affect the quality of the malt, and, therefore, the wort.

The pest yeast applies to both true spore-forming yeast and non-spore-forming yeast-like organisms. The spore-forming yeast: Saccharomyces, Picya, Hanstnula genera. The non-spore-forming ones: Candida, Torviopsis, Rhodotorula genera.

The most dangerous pests of pasteurized beer are: S.Willians, S.Fragilis, S.Diastaticus. The threshold of infection is, apparently, the presence of 4-10 cells in a bottle with the capacity of 0.33 liter. The most common of the genus Candida, C.micoderma and C.crusei, and others. The yeast of this genus is aerobe and the development of oxygen contributes to its development, that is the incomplete filling of barrels and bottles, as well as poor corking. According to Müller, the wild yeast caught in the wort at the fermentation stage cannot develop intensively.

94

But at the end of the fermentation, most of the pest yeast does not sink together with the culture yeast and develops rapidly.

On the 10th -14th day, the beer becomes turbid, there is an unpleasant smell and taste.

It has been established that the industrial yeast can also be a source of yeast infection. A dozen of wild cells in the mass of seminal yeast, with repeated use, increases its number dramatically.

Cocco-shaped bacteria of the Pediococcus genus actively develop in beer, causing a "sarcinogenic" disease of beer. The most dangerous species is the genus of Cerevisia. It forms a large amount of lactic acid. The acetic bacteria of the genera of Acetomonsa and A.Capsulatum form mucus in beer.

The bacteria of the intestinal Eisherichia group: when they multiply, the beer becomes sweetish, and acquires the smell of boiled cabbage. The presence of bacteria of the intestinal group is an indicator of the sanitary condition of the plant.

That is why, lately, great attention has been paid to the search for a quick and accurate method for detecting a yeast infection in the production of beer.
The best measures to prevent the development of wild yeast in beer production are the use of pure seed yeast, as well as maintaining the necessary biological purity at the plant.

In the 60s, the so-called killer yeast was extracted from the brewing industry, causing fast death of the culture yeast. The killer yeast was found among brewer's yeast, fermented wine, baker's, and all the wild yeast. They release the toxic substance –

"the killing factor".

It was found that the top-fermenting yeast is more sensitive to killers. At the same time, out of 140 strains of bottom-fermented brewer's yeast, 10 strains were stable, the rest were neutral.

According to Thomas, the killer yeast introduced into the bottom-fermenting seed yeast in the amount of 3% slowed down the fermentation up to 68% and impaired the beer quality. At the same time, with the continuous fermentation (top), they caused up to 80% death of the culture yeast. The killer yeast has been identified as Sacch.Carlsbergensis.

The biological purity is determined by microscopy of at least 20 fields of view with at least 50 cells in each of them. Thus, approximately 1000 yeast cells of the production culture are scanned and the content of the wild yeast is determined.

95

The yeast containing not more than 0.5% of wild yeast is considered suitable for use. Tentatively, the yeast can be evaluated according to the following scale: the number of extraneous microorganisms in 50 fields of view 0-2 - very good; 2-4 good; 4-10 - satisfactory; 10-50 bad.

Chapter III. Epidemiology

The word "epidemiology" appeared long ago, when the infectious diseases were widespread and most of them appeared in the form of epidemics. The word "epidemic" in translation into Russian means "public".

D.K. Zabolotny

The first definition of epidemiology was given by D.K. Zabolotny (1866-1929) - the founder of the school of epidemiologists.
The science of epidemics studies the causes of the emergence and development of epidemics, identifies the conditions that favor their spread and offers ways to deal with them, based on the data of science and practice.

Currently, when there is a natural decrease in the incidence of infectious diseases, and some are not recorded at all, this formulation made by D.K. Zabolotny requires clarification.

Epidemiology is a science that studies the objective laws of the emergence, spread and termination of infectious diseases in the human group, and develops measures for the prevention and control of them.

96

At the International Epidemiological Symposium in Prague (1960), the following definition was unanimously adopted: "**Epidemiology** is an independent branch of medical science that studies the causes of the occurrence and spread of infectious diseases in human society and uses the acquired knowledge to combat, prevent and, in the end, eradicate these diseases completely".

Epidemiology is divided into general and private.

General epidemiology makes it possible to study the patterns of the spread of infectious diseases among the population (the characteristics of infection sources, transmission mechanisms, susceptibility, etc.), general principles for their prevention and control.

Private epidemiology studies the epidemiological characteristics of each infectious disease individually, as well as the preventive and control measures.

If to talk about the history of epidemiology, it should be pointed out that even the ancient peoples had an idea about the contagiousness of certain diseases and tried to carry out the first preventive measures (the artificial vaccination of human smallpox), disinfection. All these measures were empirical in nature and often were ineffective or insufficiently rational.

The idea of a living causative agent of infectious diseases arose even in ancient philosophers - Hippocrates (5th -4th century BC), Lucretius (1st century BC), but it was not possible to confirm this assumption with the verified facts. Subsequently, numerous epidemics of plague, smallpox and typhoid, which were observed in the era of feudalism, especially in the 14th -15th centuries, prompted this idea. At that time, the Italian scientist Fracasco (1478-1553) came up with a formulated theory proving the contagious nature of these diseases.

In Russia, already in the 11th century, the isolation of patients and burial in special cemeteries of those who died from epidemics were used.

To prevent the plague from entering Moscow in 1552, the guard posts were first put up. In the 17th century in Russia, when the epidemics appeared in cities, in addition to the external quarantine outposts, the posting was widely practiced near the houses with sick people inside.

97

In case of the residents' death, the houses were sealed, and the things and clothes of the deceased were burned. A rule was introduced about the notification when a contagious disease appeared: "Where anyone gets sick with ulcers in the courtyard, then one should inform the sovereign."

The eighteenth century was marked by the discovery of a safe and highly effective way to prevent smallpox by vaccinating human smallpox made by Gener (1949-1823).

The activity of the first Russian epidemiologist Danila Samoilovich (1742-1805) belongs to this period. Danila Samoilovich organized the quarantine service on the Black Sea coast, participated in the military campaigns and is world famous for his works on epidemiology.

The second half of the 19th century was marked by the powerful development of physics (optics), chemistry, biology and other sciences, which contributed to the emergence of a new science, i.e. microbiology.

The ingenious scientific discoveries of L. Pasteur, I.I. Mechnikov, R. Kokh, V.P. Ivanovskii contributed to the study of not only the etiology, ptogenesis, clinic, but also the epidemiology of infectious diseases. A study of the epidemiology of a number of infectious diseases and the development of preventive measures have shown that social factors play an important role in the occurrence of epidemics.

G.N.Minh (1836-1896) and O.O. Mochutkovsky (1845-1903) made a great contribution to the development of epidemiology. They infected themselves with the blood of patients with recurrent (G.N. Minh) and rash (O.O. Mochutkovsky) typhus. With those experiments they proved that the causative agents of these diseases are in the blood of patients and their transmission is possible by blood-sucking insects.

The research on diphtheria (serotherapy), scarlet fever (studying the etiology of vaccine preparation and vaccination), and the epidemiology of malaria are associated with the name of G.N.Gabrichevsky (1860-1907). The works of E.N. Pavlovsky in the field of parasitology gained the world fame. In addition, he developed the theories of the natural foci of a number of transmissive diseases.

The development of a number of theoretical problems of epidemiology by the epidemiologist E.N. Pavlovsky and others had a beneficial effect on the practice. This is, first of all, the doctrine of the epidemiological process that arises and is supported only by the interaction of the source of infection, the specific transmission mechanism and the population susceptible to this disease

98

the doctrine of the mechanism of transmission of the infectious diseases pathogens, the doctrine of the natural foci of infections, the rational classification of infectious diseases, which is based on a significant evolutionary attribute proposed by L.V. Gromashevsky - the law of correspondence between the localization of the pathogen and the mechanism of the infection transmission. The research is continuing on the evolution of infectious diseases, the genetics and variability of germs, immunity, allergies, and other problems. A major achievement of modern epidemiology is the development of the elimination of infectious diseases. It is proved that the elimination of any infectious disease is impossible without the destruction of its pathogen as a biological species in nature.

3.1. Infection and infectious process

The term "infection" (Latin Infectio - infection and infectious process) means the totality of biological processes that occur in a macroorganism when the pathogenic microorganisms are introduced into it, regardless of whether this implies the development of an explicit or latent pathological process or it will be limited only temporary carriage.

The infectious process is a historically established interaction of a susceptible human body and a pathogenic microorganism in certain environmental and social conditions, the extreme degree of which is an infectious disease. From a biological point of view, the infectious process seems to be a kind of parasitism: two living organisms come into the struggle, adapted to the various influences of their environment.

During the infection, many parasites change their localization in the host organism or penetrate into tissues from open cavities, from the surface of the body, and the mucous membranes. In this case, the active mobility of parasites, the ability to produce various enzymes and other means of aggression play a great role.

According to the definition of E.N.Pavlovsky, parasitism is one of the forms of cohabitation in which one organism lives off an organism of another species, using it for nutrition, permanent or temporary residence.

The infectious diseases are regarded as the phenomena involving biological and social factors. So, for example, the mechanism of transmission of infectious diseases, their severity and outcome are mainly determined by the social conditions of human life.

99

The infectious diseases differ from other diseases in that they are caused by living pathogens, are characterized by infectiousness, the presence of a latent period, specific reactions of the body to the pathogen and the development of immunity.

The origin of pathogenic microbes and infectious diseases goes back centuries. As a result of changes in the genetic material, numerous types of microorganisms have formed, including pathogens, which under certain conditions can cause various human diseases. The evolutionary process in microorganisms occurs as a result of natural selection.

Cocci are believed to be the most ancient bacteria. They are found in limestones of the Protozoan era, in the coal of the Paleozoic era. In the process of evolution, some species of coccal forms acquired the ability to parasitic lifestyle. The appearance of pathogenic cocci belongs to the Permian geological period. In the sediments of the Earth's formation, deep changes in the bones of reptiles were found.

The origin of mycobacterium tuberculosis also dates back to remote times. The antiquity of the mycobacteria is evidenced by the fact that currently known species are far behind each other: warm-blooded (birds, rodents, cattle, humans) and cold-blooded (fish, snakes, turtles, frogs).

For the emergence and development of the infectious process, three links are necessary: 1) the presence of a pathogenic microbe; 2) its penetration into a susceptible macroorganism; 3) certain environmental conditions in which the interaction between the microorganism and the macroorganism occurs.

The macroorganism does not remain indifferent to the introduction of the pathogen. In the process of evolution, the human (animal) organism developed natural resistance to many pathogens: natural barriers - skin and mucous membranes, impervious to many pathogens, bactericidal properties of mucus, gastric juice, delcha, tears, activity of the reticuloendothelial system, leukocytes. In addition, the human and animal body responds by developing an immune response.

The term "immunity" (from lat. Immunitas - exemption from tribute, deliverance from anything) means the immunity of the body to infectious and non-infectious agents with genetic heterogeneity.

100

the animal and human organisms very accurately differentiates between "their own" and "foreign", which ensures protection not only against the introduction of pathogenic microorganisms, but also from foreign proteins, polysaccharides and other substances.
A branch of science that studies immunity is called immunology.

3.2. Basic mechanisms of body protection
The protective factors of the body against infectious agencies and other foreign substances are divided into:
1) non-specific resistance due to mechanical, physical and chemical, cellular, humoral, as well as physiological protective

reactions aimed at maintaining the constancy of the internal environment and restoration of impaired functions of the macroorganism;

2) innate (species or hereditary) immunity, which is the body's resistance to certain pathogenic agents, which is inherited and is inherent in a particular species;

3) acquired immunity, which is a specific defense against genetically foreign agents, carried out by the body's immune system in the form of antibody production.

The first barrier to the penetration of microbes is the outer cover. It has been established that clean skin of a healthy person is detrimental to a number of microbes. The acidic environment of sweat is associated with the presence of acetic, lactic and fatty acids in it, which have a bactericidal effect on many microorganisms. Protective functions are provided by the mucous membranes. Gastric juice has really pronounced bactericidal properties against many pathogens, especially intestinal infections and foodborne infections.

If the microbe overcomes the barrier created by the skin and mucous membranes, the lymph nodes begin to perform a protective function, in which pathogenic microbes are kept and neutralized. It is in the lymph nodes where inflammation develops, which has a detrimental effect on the causative agent of infectious diseases.

The inflammation causes an increase of the body temperature, the occurrence of acidosis and hypoxia, and the development of phagocytosis.

Phagocytosis is one of the most ancient forms of protection. It is a process of active absorption and digestion of living or killed microbes or other foreign particles that penetrated into the body by the body cels.

101

The cells capable of carrying out phagocytosis were divided into microphages and macrophages by I.I. Mechnikov. Granulocytes (neutrophils, eosinophils) are classified as microphages. Macrophages are blood monocytes, histiocytes, spleen cells, lymphatic tissue, liver cells, endothelium of blood vessels.

It should be mentioned that the totality of the anatomical and physiological characteristics of the body provides a certain degree of species stability, including the immunity to certain microorganisms.

Innate (species or hereditary) immunity

The innate immunity is the resistance of certain animal species to pathogens affecting other species. It is transmitted by inheritance from one generation to another.

An example of species immunity is the immunity of people to the plague of cattle, dogs, chicken cholera, equine infectious anemia.

Acquired immunity

Numerous observations of ancient peoples found out that people who suffered an infectious disease did not become infected when caring for a patient with that disease.

In Southeast Asia, children were vaccinated against smallpox in 2000 BC. The Iranians injected dried and powdered pox crusts into the skin incisions. Thereby, they developed immunity.

The acquired immunity is divided into natural and artificial. After having been ill with a particular infectious disease, the antibodies to this pathogen are produced in the body and a repeated penetration of this pathogen into the macroorganism does not cause the development of an infectious disease. The artificially acquired immunity is created against a particular infectious disease by introducing a weakened or killed causative agent of the disease, which causes the development of artificial active immunity, into the body.

There is a concept of the artificial passive immunity. It is formed by the ready-made antibodies which are introduced into the body along with serum.

102

Therefore, this is not the active immunity, since the body did not participate in the production of antibodies.

There is also the colostral immunity.
It is formed by the antibodies in the mother's body which are passively transmitted to the child against a particular causative agent of the disease.

In contrast to the acquired immunity, passive or active, there is the concept of the acquired immunodeficiency syndrome. This is a slowly progressive infectious disease resulting from infection with the human immunodeficiency virus, affecting the immune system, as a result of which the body becomes highly susceptible to infections and tumors, which ultimately lead to the death of the patient. Having entered the human body, the pathogen is able to initiate quite a lot of different types of differentiated cells: first of all, CD4 lymphocytes, as well as macrophage monocytes, alveolar macrophages, lymph node follicular dendritic cells, intestinal epithelial cells, cervical cells, Langerhans cells. The virus runs ahead and hits the immune system before it can even respond to a previous hit.

Immunological tolerance (lat.tolerantia - patience) - the absence of the body's immune response to a specific antigen. The immunological tolerance occurs when the body contacted with this antigen in the embryonic period. The antigens that cause immunological tolerance are called tolerogens.

The immunological tolerance can be lost as a result of the disappearance of an antigen in the body, as well as the introduction of immune serum against antigens that induced the tolerance.

Transplant immunity

The main objective of microbiology has always been to study the mechanism of immunity and reproduction of the artificial immunity. In connection with the development of surgical techniques for organ and tissue transplantation, great efforts are made to find means by which it will be possible to suppress protective reactions aimed at the destruction of transplanted grafts.

103

The success of transplantation depends on the biological compatibility of the tissues and organs of the donor and recipient, which must be genetically identical. Full compatibility of tissue cells is possible only between the identical twins.

The prevention of the development of transplant immunity can be achieved by suppressing the activity of the body's immune system.

3.3 Allergy

It has been established that the response of the immune system of a macroorganism to the substances carrying foreign information can determine the state of both immunity and hypersensitivity.

The body's reactivity, changed under the influence of pathogenic microbes, toxins, and medications, was called by C.Pirque allergy (Greek.allos - another, ergon - action). Allergy occurs as a result of a violation of the usual course of general or local reactions, more often with repeated intake of the substances called allergens.

Food allergy

Food allergy is any allergic reaction to normal, harmless food or food ingredients. Any one type of food may contain many food allergens. As a rule, these are proteins and much less often fats and carbohydrates. In allergies, the immune system produces antibodies in excess of the norm, thereby making the body so reactive that it perceives a harmless protein as if it were an infectious agent. If the immune system is not involved in the process, then this is not a food allergy, but food intolerance.

True food allergies are rare (less than two percent of the population). Most often, its cause is heredity. In children, allergies usually appear in the early years of life (often to egg whites), and then they "outgrow" it. Among the adults who think they have a food allergy, approximately 80% of them actually experience a condition that the experts call "food pseudo-allergy." Although the symptoms that are observed in them are similar to those that occur with a true food allergy. The reason may lie in the ordinary intolerance of food.

104

Moreover, some people may experience psychosomatic reactions to food because they consider food to be an allergen.

Causes of food allergies

A child whose parent is allergic has a twice higher risk of developing an allergy than the one whose parents are not allergic. If both parents suffer from allergies, the child's risk of developing it is doubled, and, thus, becomes four times greater. However, the substances that are allergens to the child may differ from the allergens of his parents.

Despite the fact that an allergy can develop on almost any type of food, the most common allergens are milk, eggs, fish, shellfish, soy, wheat and nuts, especially peanuts.

There is also a cross allergy that develops following a reaction to any one allergen. Thus, people who are allergic to peanuts can become allergic to other legumes, including peas, soybeans, and lentils. Also, patients with an allergy to cantaloupe may eventually develop a reaction to cucumbers and pumpkins; and in the same way, those who are allergic to shrimp become susceptible to crabs.

Some people develop an allergy to sulfites - chemicals used to preserve the color of food, for example, in dried fruits and vegetables. The reaction to them includes intermittent breathing or allergic shock after eating food containing sulfites. Sulfites can also cause severe asthma attacks.

Allergic reactions develop from a few minutes to 2 hours after eating. However, in a patient with a severe allergy simply touching or smelling food can trigger an allergic response.

3.4. Anaphylaxis - immediate hypersensitivity

One of the forms of the altered reactivity is anaphylaxis, a condition of increased sensitivity of the body caused by the repeated administration of foreign proteins. The first dose of antigen, which causes the increased sensitivity, is called sensitizing, the second dose, after which the anaphylaxis develops, is challenging. If you previously sensitize a guinea pig by subcutaneous injection of

105

0.01 ml of horse serum, and then after 8-21 days inject it with 0.1 - 0.5 ml of the same serum, then after 5-10 minutes the animal dies.

The anaphylactic shock in humans occurs due to repeated injections of heterogeneous immune serums in the treatment of patients with various infectious diseases (diphtheria, tetanus, anthrax, anaerobic infection), and in some cases the shock ends in death.

The allergic reactions in general and anaphylaxis in particular are very complex processes.

3.5. Delayed hypersensitivity

Allergy is a specific reaction of the body to an allergen. It develops under the influence of microbial protein, toxin, medications, pollen and other substances.

A high percentage of allergic diseases among people makes it necessary to develop such important issues as zoning of allergic diseases (establishing a geographical map of allergens), diagnosis, therapy and prevention of allergic diseases at the modern scientific level.

For the treatment and prevention of patients with allergies, the identification of allergens and the termination of contact with them (food, drugs, household, industrial and other allergens) is important.

3.6. Forms of infection and their characteristics

The forms of infection or infectious process are extremely diverse and bear various names depending on the nature of the pathogen, its localization in the macroorganism, the distribution pathways and other conditions.

The **exogenous** infection occurs as a result of infection of a person with pathogenic microorganisms coming from the environment with food, water, air, soil, secretions of a sick person and microcarrier.

The **endogenous** infection is caused by representatives of the normal microflora of conditionally pathogenic microorganisms of the individual. It often occurs with immunodeficiency conditions of the body.

106

Autoinfection is a type of endogenous infection that occurs as a result of self-infection by transferring the pathogen (usually by the hands of the patient himself) from one biotype to another. For example, from the oral cavity or nose to the wound surface. Depending on the localization of the pathogen, a focal infection is distinguished, in which microorganisms are localized in the local focus and do not spread throughout the body. For example, with furunculosis, staphylococci are in the hair follicles, with angina, streptococci are found in the tonsils, with conjunctivitis, the pathogen is localized in the conjunctiva of the eye.

However, a focal infection, with the slightest imbalance between macro-and microorganisms, can go into a generalized form in which the pathogen spreads through the body via the lymphogenous or hematogenous route. In the latter case, bacteremia or viremia develops. Blood in such cases is only a mechanical carrier of the pathogen, since the latter does not multiply in it.

The most severe generalized form of infection is sepsis. This condition is characterized by the multiplication of the pathogen in the blood with a sharp inhibition of the basic mechanisms of immunity.

In sepsis or septicemia, the main breeding ground for microorganisms is blood. Close to sepsis is the cyclic multiplication in the blood of some protozoa (malarial plasmodium, trypanosomes) with the corresponding protozoal invasions. With the occurrence of purulent foci in the internal organs, septicopyemia begins, and with the mass entry of bacteria and their toxins into the blood, bacterial or toxic-septic shock develops.

Monoinfection is caused by one type of microorganism, while mixed infection is caused by two or more types. Mixed infections are most severe. They can be caused by different bacteria: staphylococcus, proteus, Pseudomonas aeruginosa, as is often observed with nosocomial infections.

Mixed infections include many respiratory diseases caused by bacteria, viruses, mycoplasmas in various combinations.

A consequtive infection should be differed from mixed infections, in which another, caused by a new pathogen, joins the original main developed disease.

107

For example, with typhoid fever, pneumonia caused by other bacteria or viruses can occur.

Reinfection is a disease that occurred after a previous infection in the event of repeated infection with the same pathogen, for example, reinfection with dysentery, gonorrhea, syphilis, etc. In other diseases, the transmission of an infectious disease does not end with the formation of immunity.

In those cases where infection of a macroorganism by the same pathogen occurs from recovery, the superinfection occurs.

A relapse is the return of the clinical manifestations of the disease without re-infection due to the remaining pathogens in the body, for example, relapses with erysipelas, osteomyelitis, and relapsing fever.

The acute and chronic infections are distinguished by the

duration of the interaction of the pathogen with the macroorganism, as well as by clinical and pathogenetic signs.

The acute infections occur in a relatively short time. They are characterized by the pathogenesis specific to the disease, as well as clinical symptoms. In some cases, acute infections become chronic, the duration of which ranges from several months to many years.

The chronic infections are characterized by a long stay of microorganisms in the body. Depending on the properties of the pathogen and the reaction of the organism, infections can be primary and secondary, in the case of a transition of an acute infection into a chronic one. In acute infections, the pathogen usually disappears from the body soon after recovery. In chronic infections, it disappears from the body for longer periods and can be released into the environment.

The condition in which the excitation of the pathogen continues after the clinical recovery of the patient is called microcarrier (bacteriocarrier, virus carriage). Most often, these conditions are formed with weak tension of post-infectious immunity, for example, after the transfer of intestinal infections (typhoid fever, dysentery), childhood infections (diphtheria, scarlet fever). Microcarrier may also develop in healthy individuals in contact with patients or carriers of the corresponding pathogenic microorganisms. Clinically, the microcarrier does not occur.

108

In those cases when the infection proceeds without any pronounced symptoms, it is called asymptomatic, and in the presence of a characteristic symptom complex - manifest.

3.7. The dynamics of the development of the infectious process

The development of the infectious process consists of the incubation, prodromic periods, the height of the disease and the recovery period.

From the moment a pathogenic microbe is introduced until the first signs of the disease appear, a certain period of time passes, called the incubation period, the duration of which is not the same - from several hours (cholera, toxicoinfection, plague) to several months and years (leishmaniasis, leprosy). The duration of the incubation period depends on the general resistance and specific immunity of the human body, its reactivity, sensitization (increased sensitivity), the influence of harmful environmental factors and social living conditions, dose and virulence of the pathogen.

In some diseases, the incubation period begins with a prodromic period (the period of the precursors of the disease), during which there are usually no symptoms characteristic of the disease and non-specific symptoms common to many diseases develop (malaise, loss of appetite, weakness, sometimes subfebrile temperature).

During the period of the main manifestations of the disease, the infectious process, having reached the highest intensity, keeps at this level for a certain time, which is not the same for various infections.

The most typical signs of an infectious disease are fever, inflammation, damage to the central and autonomic nervous system. In addition, functional and organic disorders of the cardiovascular system, respiratory, digestive, urinary tracts are observed, and in some infections, skin changes in the form of various rashes. And the period of the disease extinction, when the course is successful, it goes into the stage of recovery, and in some cases the disease ends with a crisis - a rapid drop in temperature, accompanied by sweating and often the phenomenon of vascular collapse; and in other cases, recovery. In some cases, an infectious disease ends fatally. The bodies of infectious patients are subject to disinfection, because they contain

109

infectious agents that are dangerous when released into the external environment.

3.8.The inheritance of infectious diseases

Most researchers deny the possibility of a person having a hereditary transmission of infectious diseases caused by the infected germ cells. In addition, the transmission of infectious diseases from a sick mother to the foetus through the placenta (staphylococcal diseases, syphilis, typhoid and relapsing fever, toxoplasmosis, viral hepatitis, etc.) was precisely established.

An indisputable role in the development of infectious diseases is played by the mechanisms of the hereditary immunity, which are caused by a deficiency of the necessary substrates for the reproduction of the pathogen in the macroorganism and selective specificity for the substance of tissues and organs of people with different blood groups.

3.9. Epidemiology methods

To study the patterns of the spread of infectious diseases, the population uses a comprehensive epidemiological method, which includes epidemiological observation and experiment. In terms of prevalence, infectious diseases can be sporadic (individual diseases observed in a given area over a certain period of time). A significant excess of the level of sporadic incidence of this disease is called an epidemic (or epizootic among animals). When an epidemic reaches extremely large sizes in one country or another or covers entire countries and even continents, it is called a pandemic. In the mid-20th century, plague pandemics were repeatedly observed. Between 1817 and 1925 there were six cholera pandemics, and between1961 and 1963 the seventh pandemic of this disease began.

Over the past 60 years, viral hepatitis has taken a pandemic spread. The first pandemic occurred in 1915-1923, the second one - in 1937-1945, and the third one - in the postwar years.

In addition, there is a special form of the spread of infectious diseases - endemic, when contagious diseases persist for a long time in any locality (yellow fever, tick-borne and mosquito encephalitis, tick-borne rickettsioses, hemorrhagic fevers, leishmaniasis, tularemia, etc.).

110

In contrast to the endemic diseases, there are exotic infectious diseases that are imported from other countries.

The incidence rate of infectious disease is calculated by the number of cases during the year per 10,000 or 100,000 of the population; mortality is determined by the number of deaths from this disease per 100,000 of the population.

Taking into account the need for the wise use of the achievements of science and technology to develop more effective methods and means of combating the still numerous suffering of mankind, creating the best working and living conditions, ensuring the preservation of public health and increasing life expectancy, the 26th session of the UN General Assembly in 1972 took a decided to prohibit the development, production and stockpiling of bacteriological (biological) toxic weapon and their destruction.

3.10. Infectious disease transmission mechanism

In the process of evolution, pathogenic microbes have developed the ability in various ways to penetrate the human body and selectively localize in the tissues and organs in which they develop. For each pathogen, there are certain transmission mechanisms that determine its localization. Specific are not only the pathways of penetration of pathogenic microorganisms, but also the mechanisms of their isolation from the body. These features of the causative agents of infectious diseases are taken into account in the behavior of anti-epidemic and preventive measures.

In medical microbiology, the classification of infectious diseases is based on the etiological principle based on the specificity of the action of pathogenic microorganisms. Currently, more than 5000 agents are known that can cause human infectious diseases. Since the number of pathogenic microbial species is relatively large, it became necessary to group all infectious diseases according to a certain principle, i.e., according to the mechanism of transmission from the source of infection to a susceptible human body.

Currently, the classification proposed by L.V. Gromashevsky is adopted, according to which, depending on the location of the pathogen in the body, four mechanisms of transmission of infection are distinguished:
1) intestinal infections - fecal-oral;
2) respiratory tract infections - drip;

111

3) blood infections - vector-borne;
4) infections of the external integument - transmission of the pathogen by contact.

In addition, each group of nosological units was subdivided into two series: **anthroponoses** - diseases characteristic of humans, and zoonoses - diseases inherent in animals, but humans are also susceptible to them.

For the occurrence of the epidemic process, the source of pathogens is not enough. There are three phases of the movement of the pathogen from one organism to another:

-the first phase - the allocation of the pathogen from the infected organism to the outside;

-the second phase is his stay in the external environment;

-the third phase - the introduction of the pathogen into a new organism.

The first phase: when the pathogen is localized on the mucous membranes of the respiratory tract (influenza, measles, whooping cough), it is possible to isolate it only with exhaled air or with drops of mucus from the nasopharynx. If the pathogen is localized in the intestine, it can be excreted with feces. The presence of the pathogen in the blood is due to infection by blood-sucking arthropods.

The second phase of the transmission mechanism: the presence of the pathogen outside the body in the external environment. The pathogen isolated from the intestines enters the soil, on linen, surrounding objects, into water, and the pathogen from the respiratory system - into the air. In some cases, the transmission of the pathogen is possible without the participation of the external environment in direct contact with the patient or the carrier (sexually transmitted diseases, rabies). Air serves as a transmission factor for the so-called drip infection. After the droplets of mucus and sputum have dried, the pathogens enter the dust (tuberculous mycobacterium, variola virus, anthrax, tularemia). Water is a factor in the transmission of pathogens (cholera, typhoid fever, leptospirosis, dysentery, infectious hepatitis - these are waterborne epidemics).

Food products are of particular importance. They are the place of storage and reproduction of pathogens of infectious diseases.

Live carriers of the infectious principle are most often the biological hosts of pathogens. They actively move and contribute to the transfer of pathogens. They are usually divided into two groups: **specific** - lice, fleas, mosquitoes, mosquitoes, ticks. In their body, the pathogen multiplies or undergoes a sexual development cycle.

112

Nonspecific - transmit the pathogen in the form in which they were received. For example, flies carry the pathogens of dysentery.

Infectious disease transmission factors determine **the third phase** of the transmission mechanism - the introduction of the pathogen into the body of the next biological host.

In the epidemiological process, a susceptible organism plays an important role. Susceptibility is understood as the biological property of the tissues of a human or animal organism to be an optimal medium for the propagation of a pathogen. This property is specific and is inherited.

All people are susceptible to anthroponous diseases, but the forms of manifestation of susceptibility are different.

To zoonotic infections, the susceptibility of a person is not the same. In the defense against infectious diseases, specific immunity plays a role.

3.11. Foodborne toxic infections.

For intestinal infections, the causative agent is localized in the intestine and released into the environment with feces.

The causative agents of intestinal infections. getting into the environment with feces, urine, vomit of the patient, can cause diseases of a healthy person in the event that microorganisms penetrated through the mouth with food and water. For intestinal infections, a fecal-oral transmission mechanism is characteristic.

An increase in the incidence of intestinal infections is observed in the warm season - in summer and in the summer-autumn period. Intestinal infections include typhoid fever, paratyphoid fever, dysentery, cholera, hepatitis, poliomyelitis, brucellosis, salmonellosis, botulism.

Diseases of people with a clinical picture of poisoning due to eating meat and other animal products have been known for a long time. Many theories arose explaining the cause of deaths. However, all these theories about the causes of "meat poisoning" were unreliable. Only in 1880 O. Bolinger, having analyzed 17 outbreaks of foodborne diseases, affecting 2,400 people and having 35 deaths, found that all these cases of diseases were associated with eating meat from animals that were killed during gastroenteritis and septicemia processes.

113

But the bacterial etiology of toxic infections of a salmonella nature was first substantiated by A. Gaertner in 1888. He isolated the bacillus from the spleen of a deceased person and from the meat of a forcedly killed cow an identical bacillus, and came to the conclusion that the bacillus isolated by him is capable of forming thermostable toxic substances that cause the development of toxic infection.

Even before the discovery of Gaertner, the American microbiologist D.E.Salmon isolated a wand, which later became known as Salmonella. They are found in the intestinal canal of animals and humans, as well as in the external environment. Many of them are mobile. They can live for a long time in dust, dried feces and manure, in soil and water. For the complete neutralization of meat, seeded with salmonella, it is necessary to bring the temperature inside the piece to 80 °C and maintain it at this level for at least 10 minutes. In frozen meat, bacteria remain viable for 2-3 years, and when stored at low temperatures (+5 - 8°C) they can even multiply. In salted meat, they remain viable for 5-6 months. And with a content of 6-7% of sodium chloride in the product, they can also multiply.

In humans, salmonella cause the foodborne toxicosis. Cases of the occurrence of toxic infection are characterized by the sudden mass and simultaneous disease of people who consumed the same food. However, there is no selection of patients in the following days. There are many forms of clinical manifestation. The disease can have gastroenteric, typhoid or cholera-like, flu-like, septic forms of clinical manifestation, as well as latent bacterial carriage.

The incubation period averages 12-24 hours, but sometimes drags on to 2-3 days.

The gastroenteric form is manifested by fever, chills, nausea, vomiting, loose stools, sometimes with an admixture of blood and mucus, abdominal pain, increased thirst and headaches. The disease is especially severe when S.Typhimurium enters the human body with food.

The typhoid-like form can begin with ordinary gastroenteritis and after a temporary recovery in a few days it is manifested by signs characteristic of ordinary typhoid fever.

114

The flu-like form is more common in people with joint, muscle, rhinitis, catarrh of the upper respiratory tract and a possible upset gastrointestinal tract.

The septic form occurs in the form of septicemia or septicopyemia, in which foci are localized in the internal organs and tissues: endocarditis, pericarditis, pneumonia, cholecystitis, osteomyelitis, arthritis, abscesses.

The leading role in the occurrence of food-related salmonellosis belongs to meat and meat products. Particularly dangerous in this regard are meat and offal (liver, kidneys, etc.) from animals that have to be killed. Intravital seeding of muscle tissue and organs with salmonella occurs as a result of infection of animals with primary and secondary salmonellosis. From the point of view of food salmonella, dangerous food products include minced meat, jellies, potions, low-grade sausages, meat and liver pastes.

When chopping meat into minced meat, the histological structure of muscle tissue is disrupted, and the resulting meat juice contributes to the dispersion of salmonella throughout the whole mass of minced meat and their rapid reproduction. The same applies to pastes. Jellies and brawns contain a lot of gelatin, and low-grade sausages contain a significant amount of connective tissue.

Salmonella carriers are often waterfowl. Therefore, their eggs (ducks, geese) and meat can be a source of food-borne salmonellosis. Less toxic infections are possible when eating milk and dairy products, fish, ice cream, confectionery (cream cakes and cakes), mayonnaise, salads.

The sources of exogenous seeding can be various objects of the external environment: water, ice, containers, knives, tables, production equipment, with which they carry out the primary processing and processing of products. The participation of biological agents in the infection of products with salmonella (mouse-like rodents, flies) is also not ruled out. The contact pathway of Salmonella infection according to the animal-human scheme is not excluded. Domestic animals (dogs, cats), as well as pigs, poultry, and even pigeons play here a certain role. The contact factor of transmission according to the "man-man" scheme is a rare phenomenon and more often occurs in children.

115

Success in the fight against salmonella and their prevention is inextricably linked with the need to comprehensively strengthen measures aimed at neutralizing the sources and factors of transmission of infections, which are called upon by specialists from the medical, veterinary-sanitary and other departments.

Through the veterinary service, prevention can be ensured by the following main activities. In livestock farms and specialized livestock farms, it is necessary to observe sanitary and hygienic rules and norms for keeping and feeding animals, to carry out recreational activities, to prevent domestic and farm slaughter of cattle and poultry. An important condition is the fulfillment of the sanitary requirements in the technological processes for slaughtering livestock and poultry, the primary processing of carcasses and organs, the processing of meat and other food products, as well as the observance of the temperature regime during their transportation and storage, so that at temperatures above 4°C, salmonella can develop.

It is important to know that salmonella-seeded meat has no organoleptic signs of freshness, since bacteria are not proteolytic, but saccharolytic. Foodborne infections in humans can arise from eating completely fresh meat.

Particular attention should be paid to the methods and mode of disposal of conditionally suitable meat and the ways of its implementation.

A certain role in foodborne infections can be played by some bacteria, united by the name "conditionally pathogenic". These include Escherichia coli bacteria (CGB), which are more often responsible for foodborne diseases. They are found in the external environment, live in the intestines of humans and animals. These are Escherichia sticks with rounded ends. All conditionally pathogenic microorganisms have a relatively high resistance. At various environmental objects, they last from 10 days to 6 months. They are resistant to high concentrations of table salt, do not die at sub-zero temperatures, and are viable in tap water. They die quickly at a temperature of 68°C and above.

One of the conditions for the occurrence of toxic infections is a large contamination by food bacteria. The incubation period for toxic infection of colibacterioid etiology in humans is from 8 hours to 1 day. Cramping in the abdomen, nausea and

116

and loose multiple stools. Body temperature rises sharply to 38-39°C.

Protein-induced foodborne infections develop 8–20 hours after ingestion. Bleeding has a rapid onset, accompanied by cutting pains in the intestines, nausea, vomiting, and diarrhea. The flesh lasts 2-3 days. Sometimes 5 days. In severe cases, cyanosis, convulsions, a decline in cardiac activity, leading to death (mortality up to 2%), are observed.

The factors of transmission of the infectious principle, as in the cases of food-borne salmonellosis, may be the meat of compelled animals. A special role is given to semi-finished meat products and prepared food products, during production and storage of which the sanitary-hygienic regime was violated. For prevention, it is necessary to take measures to protect food from contamination with these bacteria, conduct thorough heat treatment and store at low plus temperatures (4-5C). The growth and development of the pathogen in food products does not change their organoleptic characteristics. Protein bacteria have proteolytic properties, and when they grow in meat, organoleptic changes in freshness occur in the appearance of specific odors. Proteus vulgaris causes the smell of mold.

If during bacteriological examination CGB is found in samples of muscle tissue or lymph nodes, then the meat is sent for processing at elevated temperatures to cooked or cooked smoked sausages.
Moreover, the temperature inside the loaf should not be lower than 75°C. When E. coli is isolated only from internal organs, the latter are boiled, and carcasses are released without restrictions. With organoleptic indicators indicative of putrefactive decomposition, meat, carcasses and internal organs are disposed of or destroyed.

3.12. Foodborne toxicosis
Staphylococcal intoxication

Staphylococcus (Stapfylococcus aureus) was discovered by R. Koch (1878), isolated from the pus of the boil by L. Pasteur (1880). Staphylococci have a spherical shape. Their diameter is 0.8-1.0 (0.5-1.5) microns. Staphylococci do not have flagella, do not produce spores, some strains form a capsule, gram-positive. Staphylococci are facultative anaerobes that produce proteolytic and sucrose enzymes.

117

They produce urease, catalase, phosphatase, form ammonia and hydrogen sulfide, ferment glucose, levulose, maltose, lactose, sucrose, lure and glycerin to form acid.

Staphylococci synthesize more than 25 proteins, toxins and pathogenicity enzymes, are characterized by hemolytic, lethal and dermonecrotic effects.

Staphylococci and streptococci are relatively resistant to drying, sodium chloride, do not die at low temperatures. Unfavorable conditions for their growth and reproduction is an acidic environment (pH 6.0 and below), a high temperature of 75 ° C is detrimental to them. Pathogenic streptococci can cause diseases of the upper respiratory tract, abscesses of the mucous membranes and skin, and staphylococci are an etiological factor in the development of septicemia processes in animals and humans, including generalized diseases - septicopyemia and septicemia.

The toxic substances produced by pathogenic staphylococci and streptococci are classified as exotoxins. They have an enteric effect, and therefore, foodborne toxicosis in humans can be caused by a toxin without the presence of microorganisms themselves. The accumulation of enterotoxins in the products is facilitated by the degree of seeding and storage time, ambient temperature, pH, as well as the association of the development of staphylococci and streptococci with some types of aeroic bacteria and molds.

The optimal conditions for the accumulation of enterotoxins in the products are the presence of carbohydrates and proteins in their composition, a temperature of 25-35°C and a pH of 6.9-7.2. At temperatures below 20 °C and pH 6.5, the production of enterotoxins slows down, and at temperatures below 15 °C and pH below 6.0 it stops. The factors contributing to the accumulation of enterotoxins in milk is its storage at the temperatures of above 12°C.

The sStaphylococcal and streptococcal enterotoxins are thermostable and are destroyed only with prolonged boiling of products. Sources of food infection with staphylococci and streptococci are diverse. One of the leading places is occupied by animals (cows, sheep), suffering from mastitis and giving milk known to be seeded by these microorganisms. Enterotoxigenic strains of staphylococci, as well as streptococci, are isolated from the carcasses and organs of animals from animals of compulsory killed sick animals.

118

Exogenous seeding is also possible during the initial processing of food products by persons suffering from pustular diseases of the skin, primarily the hands. When coughing and sneezing, staphylococci massively infect the environment, while food. The incubation period for this type of toxicosis is 2-6 hours. Clinically, toxicosis proceeds in the form of acute gastroenteritis with the following symptoms: headache, weakness, nausea and vomiting, frequent stools appear. With staphylococcal toxicosis, a rise in temperature to 38.5 °C, a decline in cardiac activity, convulsions, cyanosis of the lips, nose, limbs, weakening of vision and even loss of consciousness with a drop in blood pressure are possible. Recovery usually occurs in 1-3 days.

For the prevention of forced slaughter of sick animals, the free sale of their meat and offal is prohibited. It is forbidden to use milk obtained from patients with mastitis for food purposes.

If there are signs of stale meat and in the presence of an odor unusual for it, which does not disappear when the sample is cooked, the carcass and internal organs are disposed of or destroyed. Finished products from which staphylococci and streptococci are isolated are sent for recycling.

-Botulinum-related food toxicosis

The causative agent of botulism (lat. - sausage, botulism - poisoning with sausage poison) ...... was discovered in Holland by E. van Ermengem in 1896. It was extracted from ham, which served as a source of poisoning for 34 people, and organs of dead people.

In Western Europe, botulism is associated with the use of sausages, in America with canned vegetables, in Russia with red fish and canned mushrooms.

The causative agent of botulism is a polymorphic bacillus 4.4 to 8.6 microns long and 0.3 to 1.3 microns wide, which sometimes forms short shapes or long threads, is slightly mobile. In the external environment, it produces oval spores arranged in such a way that they give the microbe the appearance of a tennis racket (Fig. 10), gram-positive.

119

Clostridia of botulism are severe anaerobes. The optimum growth temperature is 25-37°C, it is grown in ordinary media, pH 7.3 - 7.6, preferably in meat or brain gruel. cultures emit a pungent smell of rancid oil.

Clostridia of botulism break down tissue pieces and egg white in a liquid medium, form hydrogen sulfide ammonia, volatile amines, alcohol, acetic, butyric and lactic acids. Botulinuses produce exotoxin (neurotoxin). The content of sodium chloride in a concentration of 6-8% prevents the multiplication of the microbe and the accumulation of toxin, the lethal dose of which for humans is 0.0001 mg. Unlike tetanus and diphtheria toxin, it is absorbed unchanged in the gastric juice.
Exotoxin is activated by trypsin, as a result of which its biological activity in the intestine increases many times.

Vegetative forms of the pathogen die at a temperature of 80C in 30 minutes, spores can withstand boiling for up to 6 hours. At a temperature of 115 °C it dies in the course of 5-40 minutes, at 120 °C - in 3-22 minutes. When boiling, the exotoxin is destroyed within 10 minutes. The causative agent is found in soil, leaves, grass, hay, vegetables, fruits, on the surface of the body and in the intestines of large fish, in the intestinal canal of animals and humans, in manure.

Pathogenesis of infection in humans. The cause of the poisoning is the consumption of meat products, canned vegetables and fish, sausages, ham, salted and smoked red fish, chickens, ducks and other products infected with the botulinum pathogen.

In the presence of anaerobic conditions in food products of animal or vegetable origin, the pathogen produces a toxin.

The incubation period is 12-24 hours. It depends on the amount of toxin ingested. The shorter the incubation period, the harder the disease. The botulinum toxin acts on the autonomic nervous system, on the nuclei of the cells of the medulla oblongata and on the cells of the ganglia of the spinal cord, but does not affect the centers of the cerebral hemispheres. At the onset of the disease, patients develop general weakness, headache, nausea, sometimes vomiting, but constipation and flatulence are characteristic. In the future, there is a double vision of objects, unevenness of the pupil, dry mouth and throat, stiffness of the tongue, hoarse voice, later respiratory distress, muscle weakness. Temperature 35.5-36°C. Before death, the severe shortness of breath appears with a rib type of breathing.

120

Death occurs with respiratory paralysis, most often on the 4-8th day, but sometimes after 8-24 hours. Mortality, if treatment is not carried out in a timely manner, can be 15-17%. Antibotulinic serum is used for treatment.

The danger of poisoning is represented by canned meat and meat and vegetable products, when the raw materials for their production are contaminated with soil and intestinal contents, and the technology of their manufacture was disrupted. Of the various canned food, mushrooms rolled up in cans are especially dangerous, during the processing of which it is not always possible to wash them from particles of the earth. Poisoning can occur from poor-quality salted fish, especially from the family of sturgeon and salmon, from ham-piece products (ham, loin, brisket). Most often, botulism in humans is a consequence of the use of canned foods prepared at home with non-compliance with sanitary requirements and temperature conditions.

For the prevention of botulism, the initial processing, storage and transportation of carcasses must be carried out in compliance with veterinary and sanitary requirements. Not to allow fish and meat of dubious freshness to salting and smoking. The shelf life of the products is important. The optimum storage temperature is not higher than 3-4°C and their heat treatment before use.

Mycotoxicoses are low molecular weight secondary metabolites produced by microscopic mold fungi. Most fungi are aerobic organisms. A common feature of all mycotoxins is toxicity to animals, but also to humans. These are widespread pollutants of food raw materials and human food. In the mid-19th century, it was found that the microscopic ergot fungus that pollutes crops is the cause of the disease, known as the "fire of St. Anthony" or ergotism. The disease was observed in Spain, France, Switzerland, Prussia, Hungary in the 6-19th centuries, in Russia in the 19 century. In Russia, the last major epidemic was observed in some areas of the south in 1926-1927. St. Anthony was considered the patron saint of victims of ergotism, and the Order of St. Anthony was involved in the treatment of these patients.

121

The disease manifests itself in two clinical forms - gangrenous and convulsive, the evil cramps. In the generalized form, acute pains appear and dry gangrene develops in the limbs (up to the rejection of tissues or entire limbs). The convulsive form is most pronounced, characterized by mental disorders. Nausea, vomiting, and diarrhea are noted. Epileptic seizures are possible.

Food mycotoxicoses are characterized by foci, they occur suddenly, are not contagious. Toxic substances of mushrooms are resistant to high temperatures of 200°C and more.

Fusariotoxicosis - known as poisoning with "drunken bread" - arises from eating bread made from grain affected by Fusarium fungus. At the same time, there is a sharp excitement (unreasonable laughter, dance, singing), a shaky gait. A picture of severe intoxication appears. There are no fatal cases. The number of leukocytes is reduced to 1000 or less in 1 mm3. The number of red blood cells increases, which serves as the earliest indicator of toxicosis.

The toxicity of substances formed by Fusarium fungi is a complex of chemical compounds, of which toxic styrene - lipotoxol - plays the leading role in intoxication. They are resistant to storage, do not break down in products when baking bread, cooking porridge and fermentation.

The incubation period depends on the degree of intoxication and ranges from several minutes to several hours after eating foods prepared from infected grains.

Kashin-Beck disease. The disease was first detected in 1860 in Siberia in the Valley of the Urov river. Endemic joint disease with a violation of the processes of ossification, growth, premature wear of the osteoarticular apparatus. Deforming osteoarthrosis develops.

Imperfect fungi include a large group of dermatomycetes, of which the causative agents of trichophytosis, microsporia, and epidermophytosis are of the greatest importance. Deomatomycetes do not produce exotoxins. They contain endotoxin and are allergens that cause a state of hypersensitivity of the body and especially the skin.

122

Trichophytosis is characterized by the damage to the scalp, as well as the skin of hands and nails. Hair breaks off near the surface of the skin, and in the follicles, their residues are noticeable. The fungus is located both inside and on the surface of the hair. Pinkish-red scaly spots appear on the skin. When microsporia affects hair, skin, less often nails. The fungus penetrates the hair and is located throughout its length.

Epidermophytosis is manifested by damage to the skin, especially the interdigital areas, feet, hands, nails, the hair is not affected.

Aflotoxicosis. The causative agent of the Aspergillus genus belongs to the class of ascomycetic fungi. The pathogenic and opportunistic species include A.fumigatus, A.Flavus, A.Niger. More than 40 species have been described, isolated from patients with various clinical forms of aspergillosis. In humans, they cause damage to the skin of the trunk and limbs, adnexal cavities of the nose, lungs, bronchi, cornea of the eye, external auditory meatus, sometimes bones and other organs and tissues. Some types of pathogenic molds produce aflotoxin, possessing high toxicity and carcinogenic effect.

Food poisoning manifests itself as a result of ingestion of aflotoxins. In addition to peanuts, this fungus affects: wheat, corn, soy, rice, peas, cocoa beans and coffee. In 1960, over 100,000 turkey poults died in England (liver damage and rapid death). The reason was the Brazil nut (peanuts). The optimal conditions for the growth and development of mushrooms are: 20-30°C, humidity 85-90%. The lethal dose for humans is 2 mg per 1 kg of body weight. In this case, membrane permeability is violated, DNA and RNA synthesis is suppressed.

Carcinogenic, mutagenic and embryotoxic activity develops. In chronic intoxication, cirrhosis and primary cancer develops, in acute - fatty infiltration occurs with a focus of necrosis, renal and intestinal function is impaired.

The Penicillium genus belongs to the class of ascomycetes. This genus of the fungus is widespread in nature. It is found in feed, dairy products, on wet items, old skins, jam.

More than 30 species of Penicillium are pathogenic to humans. They cause penicillosis - damage to the skin, nails, ear, upper respiratory tract and lungs, as well as a generalized infection with the formation of foci in the internal organs.

123

Pathogenic species include P.Crustaceum, P.Notatum, P.Mycotomagenum and others.

3.13. Infectious diseases

Infectious diseases of animals by the degree of their danger to humans, the sanitary examination divides into three groups: group 1 - infectious diseases transmitted to humans through milk, meat and other slaughter products (tuberculosis, brucellosis, anthrax, leptospirosis, swine erysipelas, foot and mouth disease). Group 2 - infectious diseases that a person is sick with, but which are not transmitted through milk, meat and other products of slaughter (tetanus, rabies, actinomycosis, pseudotuberculosis, etc.). Group 3 - infectious diseases that a person does not have.

There is a fundamental difference between food poisoning and foodborne infectious diseases. The first occur when live microbes enter the food, multiply plentifully in it and enter the human body.
The second - when even a few pieces of microbes are present

in food. The penetration of one stick of anthrax into the body, multiplying intensively, can cause death.

Typhoid fever (Greek: Typhos - fog). This is a severe acute infectious disease. It is characterized by deep general intoxication, bacteremia, specific damage to the lymphatic apparatus of the small intestine. Intoxication is manifested by severe headache, blurred consciousness, delirium. The causative agent of typhoid fever is pathogenic microorganisms of the genus ..., which include a large group of bacteria, but only three of them, i.e.S.Ttphi, S.Paratyphi A and S.Paratyphi.

The primary location of pathogens is the digestive tract. Infection occurs through the mouth. After the introduction of typhoid salmonella into the intestines of a person, a certain period of time passes during which inflammatory phenomena develop in the body. At the end of the incubation period, salmonella from the lymphatic apparatus of the small intestine penetrates into the thoracic duct, and then into the blood. A state of bacteremia develops, in which hematogenous introduction of the pathogen into the lymph nodes, spleen, bone marrow, liver and other organs occurs.

124

The causative agents of these infectious diseases have much in common. Once in the gastrointestinal tract, they multiply intensively. The dispute does not form. Warming at 56 °C for 10 minutes kills the bacterium. At the same time, freezing in ice suffers for several months.

Typhoid bacilli form a potent, heat-resistant endotoxin. The only preserver of the causative agents of typhoid fever is a sick person or a bacteriocarrier. For intestinal infections, a fecal-oral transmission mechanism is characteristic. Food workers can be carriers of intestinal infections. Therefore, it is important to exercise sanitary and epidemiological control over catering workers. The pathological process proceeds in the small intestine.

It is necessary to remember that the contamination of food products, including milk, which is a good breeding ground for microorganisms, usually occurs through contaminated hands of bacterial carriers, as well as through flies. Dairy epidemics are characterized by a large number of diseases among children. Optimal for bacterial growth is a temperature of 37°C.

The most important ideological feature of the causative agents of typhoid and paratyphoid A and B is their ability to resist phagocytosis. The largest epidemics were caused by tap water pathogens. The incubation period for typhoid fever lasts 25 days, but can vary from 7 to 25 days, which depends on the infectious dose, the virulence of the pathogen and the resistance of the macroorganism.

Measures to combat these diseases include: neutralizing the source of the disease; the intersection of their distribution paths. In this regard, food industry enterprises must adhere to all sanitary requirements that ensure the prevention of food contamination by disease promoters.

Dysentery. This is one of the most acute intestinal diseases, which is characterized by general intoxication of the body, diarrhea and a peculiar lesion of the mucous membrane of the large intestine.

The disease has been known since its description by doctors of the Middle Ages as "biting diarrhea." Separate information about deserterium dates back to ancient times. In the 16th and 17th centuries, epidemics of acute disinfection repeatedly occurred.

125

The causative agents of dysentery is a large group of bacteria united in the Shigella genus, which includes 40 serotypes. The optimum temperature for bacterial growth is 37°C. Pathogens are highly resistant to environmental factors. They survive on cotton and paper for 30 - 36 days, in soil - up to 3 - 4 months, in water - 0.5 - 3 months, on fruits and vegetables - up to 2 weeks, in milk and dairy products - up to several weeks.

In the course of their life, shegelles accumulate toxins inside microbial cells, and when the pathogen dies in the body of people with dysentery, these endotoxins cause general intoxication of the body and damage to the mucous membrane of the large intestine.

In the external environment, Shigella are excreted in the feces of patients with acute, protracted and chronic dysentery, as well as bacterial carriers. In the external environment, they persist for a long time. So, for example, shigella in milk lasts up to 2 weeks, in fermented milk products - 6-8 days.

The most important biological property of shigella, which determines their pathogenicity, is the ability to invade epithelial cells, multiply in them and cause their death.

The incubation period for dysentery lasts 2-5 days, but sometimes it can be very short - 12-24 hours. First, pains appear in the abdomen, after a few hours, mucus appears in the feces, then blood streaks. Usually, all the symptoms of dysentery reach their full development in about 3-4 days.

After hospitalization of the patient, the final disinfection of the premises in which he was located is performed. An epidemiological examination of each case of the disease is carried out in order to identify the source of infection and its distribution. For canteens, food warehouses, shops and other food items establish strict sanitary control. They carry out systematic measures to combat flies. Of great importance is the control of water sources. Food industry workers should keep in mind that vegetables and fruits consumed unwashed contribute to the spread of dysentery. When heated to 60 °C, the pathogen dies after 10-15 minutes. But in food, it remains viable for up to 15 days.

Amoebic dysentery

126

A long-lasting human arazitic disease that develops when infected with an alimentary pathway, a histolytic amoeba, is transmitted both from patients and from carriers of amoeba cysts. The causative agent is Entamoeba histclytica. It exists in two forms: a) vegetative, b) encysted. In the presence of active manifestations of this disease, the feces of patients contain vegetative forms of a histolytic amoeba, which are usually divided into a luminal form that parasitizes in the upper colon, and a tissue form that invades the mucous membrane and submucous layer of the wall of the ascending part of the colon, and often and in its other departments. Penetrating into the thickness of the colon wall, the histolytic amoeba causes deep ulcers here, which are connected by passages that are formed due to the melting of tissues by proteolytic enzymes of the parasite. The greatest number of luminal forms of histolytic amoeba is found in the contents of the cecum; the luminal form has an active movement and forms proteolpitic enzymes that cause the tissue to melt in the walls of the colon with the formation of deep ulcers. Feces of patients contain vegetative forms of histolytic amoeba. Histolytic amoeba have a rounded shape and contain 4 nuclei and a vacuole located in the protoplasm filled with glycogen (Fig. 14).

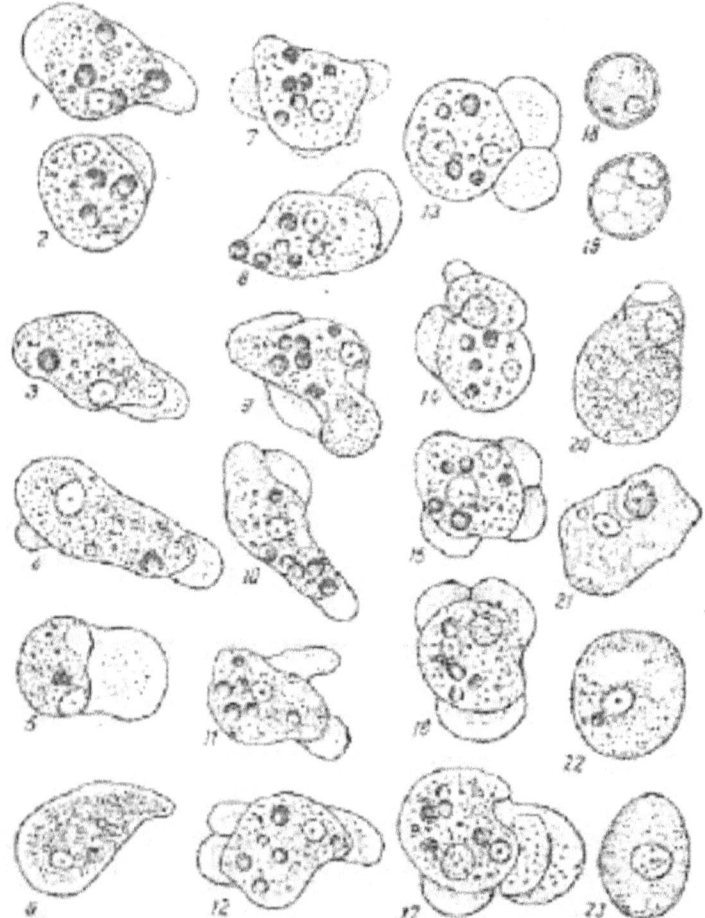

Fig. 14. Histolytic amoeba in different
(1-23) forms of development.

127

The disease spreads in areas with a warm and hot climate. Separate cases of the disease were noted in the Caucasus, as well as in areas located along the lower reaches of the Volga River and some areas of Central Asia. Reservoirs, as well as sources of infection, are patients with a subacute and chronic course of amoebiasis, in the stage of exacerbation of the disease, as well as cystocarriers. People become infected with amoebiasis as a result of the pathogen entering the gastrointestinal tract.

The cause of human infection is often drinking raw water. In addition, the cause of infection is the use of products infected with histolitic amoeba cysts. There is a certain danger of vegetables and berries that were removed from gardens that were fertilized with unreacted feces of people, among whom were patients with amoebiasis and cystocarrier.

Cholera - an acute intestinal disease with a high mortality rate

This disease has been known to mankind since ancient times. In the history of cholera, cholera pandemics are especially characteristic. During the period from 1817 to 1925, there were 6 cholera pandemics, that is, 6 times cholera captured Asia, Europe, Africa, sometimes America and even Australia. Russia - the first of the European countries where cholera entered - has survived 57 cholera years. 5.6 million people were ill with cholera; 2.14 younger people died from it.

The causative agent strict aerobic was discovered in 1883 by R. Koch in the feces of patients with cholera. Vibrio cholera is the causative agent of human cholera. It has the shape of a curved stick 1.5-3.0 μm long (Fig. 11), very mobile. Spore and capsules do not form.

Under the influence of physical, chemical and biological factors, cholera vibrios are subject to variability. In sophisticated environments and in old cultures, they can take the form of grains, balls, bulbs, sticks, threads, spirals.

Vibrio cholerae dilute coagulated serum, gelatin, form indole, ammonia. Widespread, severe course of the disease and high mortality make cholera a particularly dangerous infectious disease. Propagating in the intestines, the pathogen causes necrotization of the intestinal mucosa epithelium. The causative agent synthesizes exotoxin cholerogen, which determines the pathogenesis of cholera. It survives well at low temperatures, it remains viable in ice for up to 1 month, and in sea water for up to 47 days.

128

Cholera vibrio dies at 80 °C in 5 minutes, at 100 °C - instantly. It is highly sensitive to acids and chlorine.

The incubation period lasts from several hours to 5 days, on average 2-3 days. The pains in the abdomen, copious loose stools, wastes intestinal epithelium are marked. Vomiting appears, the patient experiences a strong thirst, general weakness. In the future, dehydration, cyanosis, the disappearance of the pulse occurs. The mortality is very high.

Of the general sanitary measures, first of all, systematic control over water sources and water supply is ensured. Sanitary supervision of the trade in products, especially consumption in the form of milk and drinks, is carried out. A systematic fight against flies is being conducted. When the water contains 1 mg of active chlorine, the pathogen loses its activity.

Brucellosis

This kind of disease of people and animals is caused by bacteria of the Brucella genus. The causative agent of brucella was discovered in 1886 by D. Bruce in the spleen of a dead person. Subsequently, cattle pathogens B.Abortus and B.Melitensis, those of pigs – B.suis - were discovered.

Brucella does not produce exotoxin. As a result of the breakdown of bacterial bodies, endotoxin is formed which has specific allergenic properties. Brucella has a common antigen with the tularemia bacterium and cholera vibrio.

Brucella is characterized by great stability and vitality. They are stored for a long time at low temperature. In soil, urine, animal feces, manure, hay dust, bran and brinze - up to 4 months, in sheep's wool - up to 3-4 months, in meat - 20 days. From the action of temperature 60°C they die within 30 minutes, at 70°C - after 10 minutes, at 80-90°C - after 5 minutes. The best results are obtained using 1% hydrochloric acid in combination with 8% sodium chloride.

The main carriers of brucellosis infection are: sheep, goats, cattle, pigs. A person is most susceptible to infection of B.melitensis. The entrance gate to the penetration of brucella into the human body is the mucous membrane of the digestive tract. In the future, the pathogen penetrates the lymphatic system, and from there into the circulatory system.

129

The incubation period is from 1 to 3 weeks. First, the temperature rises (38-39°C). Brucellosis occurs as chroniosepsis. The clinic of brucellosis is diverse and complex. Most often, the lymphatic, vascular, nervous and especially the musculoskeletal system suffer (Fig. 15).

One is ill with brucellosis for a long time, sometimes up to 10 months. In severe cases, it can lead to prolonged disability and temporary disability. However, as a rule, it ends in recovery. The disease lasts several years. You should know that the pathogen in milk, dairy products and meat remains alive from 10 to 45 days. One of the conditions for the prevention of brucellosis is careful monitoring of the manufacture of dairy products, especially feta cheese.

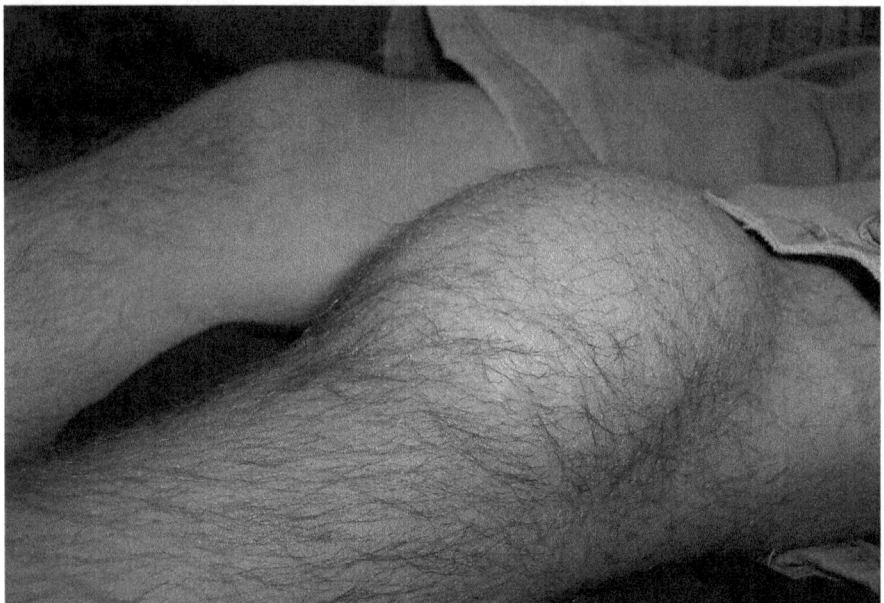

Fig. 15: Brucellosis bursitis of the left knee joint

Tuberculosis (from the Latin Tuberculum - tubercle) is an infectious disease of humans and animals, in which the formation of small tubercles occurs. Tuberculosis is ubiquitous. In the incidence of tuberculosis and its spread, social and living conditions of life are crucial, since both innate resistance and immunity acquired to it are determined by these conditions.

Mycobacterium tuberculosis - thin straight or slightly curved sticks 1-4 microns long and 0.3-0.6 microns wide. They are motionless (Fig. 16).

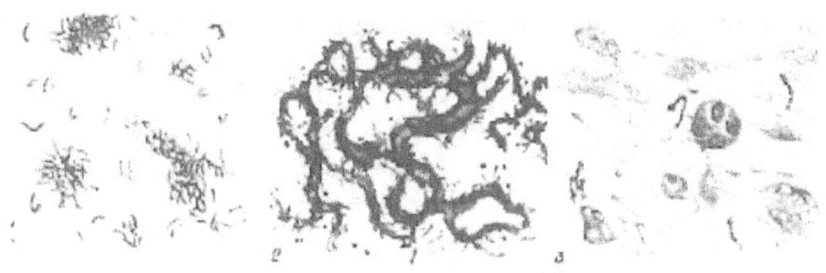

130

Fig. 16 Mycobacterium tuberculosis

1-in pure culture; 2-in culture from a liquid nutrient medium (arrangement in the form of braids); 3-in sputum (coloring according to Zil-Nielsen)

Mycobacterium tuberculosis (Mycobacterium tuberculosis) was discovered in 1882 by R. Koch. He also studied the pathogenesis and immunity of this disease. A major event should be considered the receipt of a live vaccine against tuberculosis, with which it became possible widespread specific prevention of this disease. The use of antibiotics and other anti-tuberculosis drugs has equipped modern medicine with a powerful tool in the fight against tuberculosis. The causative agent is motionless, does not form a dispute. The mycobacterium tuberculosis - aerobes, optimum growth 37°C. The causative agent grows in elective environments.

Proteolytic enzymes were discovered in tuberculosis microbacteria that break down protein in an alkaline and acidic environment; they ferment alcohol, glycerin and numerous carbohydrates, lecithin, phosphatides, urea.

Tuberculous mycobacteria do not produce exotoxin. They contain toxic substances released upon cell breakdown.
In 1890, R. Koch received tuberculin from tuberculosis bacteria. In the pathogenesis of tuberculosis, fatty acids (butyric, palmitic, tuberculostearic, aleic) are of great importance; they contribute to the cheesy degeneration, decay of cellular elements, cause the blockade of enzymes (lipases, proteases).

It should be noted that tuberculin, first proposed by R. Koch, is widely used to reproduce an allergic test, by which infection with tuberculous mycobacteria is determined.

Compared to other non-spore forming bacteria, tuberculosis

microorganisms are more resistant to external factors. Mycobacterium tuberculosis lasts for more than a year in running water, up to 6 months in soil and manure, over 3 months on book pages, 2 months in dried sputum, and several weeks in distilled water. They are easily neutralized at a temperature of 100-120°C; sensitive to sunlight.

Tuberculosis is an infection widespread among cattle, chickens, turkeys, and others; pigs and small cattle are less likely to get sick.

131

It has been established that human tuberculosis is caused by two main types of bacteria - human (M.Tuberculosus) and bovine (M.Bovis). Tuberculosis can be caused by infection and by bird species.

The infection with tuberculosis occurs by airborne droplets and airborne dust, sometimes through the mouth when eating foods infected with tuberculous mycobacteria, through the skin and mucous membranes; intrauterine infection of the foetus through the placenta is also possible.

With aerogenic infection, the primary infectious focus develops in the lungs, and with alimentary infection, in the mesenteric lymph nodes.

The source of infection is a person with tuberculosis, less commonly animals.

From a sick person, the agent is released mainly with sputum, as well as urine, defecation and purulence.

The tuberculosis bacillus can affect almost any organ and any tissue.

The human body has a high natural resistance to the causative agent of tuberculosis. In most cases, the primary infection does not lead to the development of the disease, but to the formation of a focus, its delimitation and calcification.

It has been established that of all infected people (about 80% of the adult population over 20 years old) no more than 10% get sick, and only 5% directly after infection. The immunity with tuberculosis is non-sterile.

With tuberculosis, an indisputable role is played by a genetic factor in immunity, which has been studied in detail in twins. In identical twins, the incidence concordance in the incidence is 67%, in twin twins - 25.6%, in brothers and sisters - 25.6%.

Tuberculosis treatment is carried out with the help of antibiotics and chemotherapy. Prevention of tuberculosis involves the implementation of broad socio-economic measures, early and timely identification of patients with tuberculosis and the provision of effective medical care.

One of the protection factors should be called phages that have an effect on both virulent and avirulent strains of tuberculous mycobacteria. The discovery of phages has a certain practical significance.

132

They can be used in the diagnosis of tuberculosis and, possibly, in its therapy.

The prevention is provided through early diagnosis, timely detection of patients with atypical forms, medical examination. the neutralization of milk and meat from animal animals and social activities (improving the working and living conditions of the population, raising its material and cultural level).

Active immunization of people is very important in the fight against tuberculosis, which causes a significant decrease in morbidity, a decrease in the severity of the disease and mortality, reduces the sensitivity of the body to the action of tuberculous mycobacteria and their degradation products, stimulates the body to fix and neutralize the pathogen, increases the biochemical activity of tissues and causes more enhanced production of antibacterial substances.

In this country, the method of intradermal immunization and revaccination is used. For this purpose, a special dry BCG vaccine is produced, which is administered to newborns once on the outer surface of the left shoulder. Revaccination is performed at the age of 7, 12, 17, 23 and 27-30 years. Post-vaccination immunity develops after 3-4 weeks and lasts from 1 - 1.5 to 15 years.

A greater role is played by living conditions, violation of which increases morbidity and mortality (wars, famine, unemployment, economic crises and other disasters). According to WHO, a high incidence of tuberculosis is observed in Latin America, India and Africa.

Yersiniosis. This is a zoonose acute bacterial infectious disease with a fecal-oral pathogen transmission mechanism.

In the genus yersinia includes three types of bacteria (y.Pestis, Y.Pseudotuberculosis, Y.Enterocolitica). The causative agent Y.Enterocolitica is an ovoid and rod-shaped cell, does not form capsules, grows well on ordinary media; optimal growth temperature 30-37°C. The causative agent in humans causes gastroenteritis with damage to the intestinal walls, with fever, with symptoms of intoxication.

The source of infection of yersiniosis are various species of animals, the main specimen of pigs and cattle, as well as dogs, cats, rodents, birds. Infected animals and sick people excrete the pathogen with feces and urine, polluting water, plants and other objects of the environment through which the person becomes infected.

133

The infection occurs as a result of eating raw or insufficiently thermally processed foods: meat, meat products, milk, vegetables, fruits, greens. The causative agent is able to multiply both on plants and inside them, for example, in lettuce, peas, etc.

Intestinal yersiniosis is found in all countries of the world, but the incidence rate is higher in economically developed countries with a developed network of centralized food supply. Group diseases are more often associated with the use of various vegetable salads, especially from cabbage, which was stored in contaminated rodent secretions in vegetable stores.

The disease is characterized by acute onset (incubation period of 3 - 7 days), fever, intoxication, abdominal pain, diarrhea, inflammation of the skin, pain in the joints and muscles.
Prevention is associated with the prevention of the penetration of rodents into vegetable stores, food warehouses, shops, catering establishments.

Anthrax is an acute infectious disease of humans and animals, the causative agent of which is a bacterium of the species Bacillus anthracis. In Russia, this disease was called anthrax in connection with the large epidemic described in the Urals in 1786-1788 by S.S.Andreevsky.

The siberian ulcer bacilli are large: their length is 3-5 microns, width 1-1.2 microns. They are arranged in pairs or short chains in the body and long chains on nutrient media.

Bacilli are immobile, outside the body they form spores of an oval shape, located centrally, but exceeding the diameter of the cell.
It has been established that under favorable conditions, spores in the warm season can grow into vegetative forms and, with the onset of autumn, become spores again.

Anthrax bacilli in the body of animals and humans form capsules surrounding both individual individuals and chains.

One of the important factors of the causative agent of anthrax is the capsule. protecting bacteria from phagocytosis. Loss of capsule leads to a loss of virulence.

134

Another major virulence factor is the complex toxin complex. Anthrax bacillus, when grown on a semi-synthetic medium, secretes an exotoxin protein complex containing a lethal toxin into the culture fluid.

Siberian ulcerated bacilli in a bouillon culture in sealed ampoules mature up to 40, and spores up to 65 years. In the dry state, spores remain alive until 28 years old, in the soil, according to some sources, up to 100 years. Spores can withstand boiling for 15-20 minutes from autoclaving at 110°C, they die within 5 minutes.

From domestic animals are susceptible sheep, cows, horses, deer, camels and pigs.

Anthrax is a typical zoonotic disease. People become infected from sick animals, as well as through objects and products from infected raw materials (short fur coats, fur gloves, collars, hats, brushes for shaving, etc.). In the summer, infection is possible through bloodsucking insects. The disease manifests itself in three main clinical forms: cutaneous, pulmonary and intestinal.

With a skin form, the pathogen penetration site is damaged skin, mainly exposed parts of the body (face, neck, hands, forearms).

In the localization site of the causative agent, an anthrax carbuncle is formed.

In the pulmonary form, infection occurs aerogenically during work with materials infected with spores of anthrax bacilli. The disease proceeds as a type of severe bronchopneumonia. The bacilli excreted in sputum.

The intestinal form occurs as a result of eating meat from sick animals. In this case, a severe lesion of the intestinal mucosa with hemorrhages and foci of necrosis is noted. Bacilli stand out with bowel movements. With this form, general intoxication is observed. The disease lasts 2-4 days and most often ends in death.

135

The comprehensive treatment of anthrax patients is carried out, directed both against toxin and bacilli, timely intramuscular administration of antisyphyrales globulin (30-50 mg) and antibiotics.

The bodies of animals killed by anthrax are burnt or buried in a designated place (cattle cemetery) to a depth of at least 2 meters and covered with bleach.

Vaccination of those individuals who have the possibility of infection with anthrax is practiced. Currently, in Russia, for the prevention of anthrax in humans and animals, a highly effective live spore capsule-free SMT vaccine is used, which is prepared from an avirulent strain of the bacillus Bac.Anthracis.

The phenomenon of immunodeficiency is due to a violation of the immune status, a defect in one or more mechanisms of the immune response. There are primary, or congenital, and secondary, or acquired, immunodeficiencies. The primary immunodeficiencies are conditions. in which a violation of the immune mechanisms is caused by genetic defects. Humoral, cellular and combined immunodeficiencies are distinguished

187

depending on the level of violations and the localization of the defect. They may be associated with phagocytosis deficiency. Congenital immunodeficiency syndromes and diseases are quite rare. Their causes are damage to the genome in the embryonic period.

The clinical picture manifests itself in the form of infectious complications, allergic reactions, congenital malformations.
The insufficiency of humoral immunity is a violation of the synthesis of immunoglobulins or accelerated decay.

The lack of cellular immunity is due to a violation of the functional activity of T cells, since T lymphocytes affect the functional activity of B cells. Children with T-cell immunity die at an early age from serious infections or malignant tumors. **Combined** immunodeficiencies develop with a combination of disorders of the T and B parts of the immune system.

The secondary immunodeficiencies are observed in individuals with an immune system that has functioned normally from birth and is accounted for by its damage after infections and invasions, burn disease, uremia, and tumors; metabolic and invasive disorders, with burn

136

with burn disease, uremia, tumors; metabolic disorders and depletion as a result of drug exposure.
Secondary immunodeficiencies include AIDS. The human immunodeficiency virus (HIV) belongs to the family of retroviruses. The source of the causative agent of the infection is an infected person who is at any stage of infection, including during the incubation period. The virus can be found in all biological fluids (blood, semen, vaginal secretion, breast milk, saliva, tears, sweat, etc.). The dominant mechanism of transmission of the pathogen is contact, implemented through sexual contact. A high concentration of the virus in semen and

vaginal secretion was noted.

Listeriosis. This is a zoonotic bacterial infectious disease with a fecal-oral pathogen transmission mechanism. The causative agent of the disease (Listeria monocytogenes) was discovered in 1926. E. Marey and named in 1940. G. Peary in honor of D. Lister. Listeria are small with rounded ends cocciform bacteria with a length of 0.5 - 2 microns, motile, with polar flagella, gram-positive. They are located singly or in pairs.

Listeria are facultative anaerobes, unpretentious, develop at 37C on ordinary sections. They secrete thermolabile hemolysin into the culture fluid, which, as a result of its activation with cysteine, causes hemolysis of the red blood cells of pigeon, rabbit, and horse. Upon decay, endotoxin is released from microbial cells, which causes changes in animals and humans.

Listeria are resistant to environmental factors. In the dried state, they retain their pathogenic properties for 7 years. They are resistant to freezing. They can withstand the temperature of 55°C for 1 hour. They die from boiling after 3 minutes, 1% and a 0.5% formalin solution are detrimental to them.

In vivo, cattle, horses, pigs, rabbits, hens, pigeons are affected by listeriosis. A person becomes infected from sick rodents, pigs, horses. The most dangerous for humans are meat products from pigs suffering from listeriosis. Infection is possible with a tick bite in the endosootic organs of listeriosis. The causative agent enters the body through damaged integuments and mucous membranes of the mouth, nasopharynx, conjunctiva and digestive tract. The disease is characterized by sepsis (acute and chronic), the phenomenon of meningoencephalitis, which in a large percentage of cases ends in death

137

especially among newborns and the wounded in the brain.
In the throat, inflammatory processes are noted, sometimes a rash appears on the skin.

In the pathogenesis of listeriosis, they attach importance to the whole organism or individual tissues and organs by the endotoxins of the pathogen, which multiplies intensively in the body of infected animals and people.

The lymph nodes, central nervous system and internal organs of newborns are most affected.
In humans, the main forms of listeriosis are religious-septic and nervous. The first usually ends in recovery. in a number of cases, the second form causes death.

In animals that have undergone listeriosis, immunity is developed. In humans, immunity is not well understood, hyperimmunach serum does not have a healing property.

The prevention is ensured by carrying out joint sanitary and hygienic measures together with the veterinary service, laboratory control of meat sold for sale. Ensure the burial of the bodies of dead animals. To carry out disinfection and deratization in the outbreak; compliance with personal hygiene, especially working in the food industry; milk consumption only after boiling, and meat - only after thorough heat treatment.

Leptospirosis is a zoonotic bacterial infectious disease that is caused by Leptpspira interrogans. Leptospira are microorganisms with 12-18 small primary curls and resemble a dense spring, they are mobile, make rotational and translational movements. In figure 17, leptospira is presented in an electron microscopic version.

138

Fig. 7. Leptospira in an electron microscope

In morphological terms, pathogenic and saprophytic leptospira are indistinguishable. Under the influence of a nutrient medium, elevated temperature and prolonged cultivation, atypical forms can be observed in leptospira.

Leptospira do not produce exotoxin. Toxic substances are found only in living leptospira parasitizing in humans and animals. They are quite resistant to the influence of external factors. 5-10 days survive in water, in the soil - a little longer. In food products (milk, butter, bread, etc.), the viability of leptospira does not exceed several hours. Leptospira persist for a long time at a low temperature (-70 - 90°C).

Under natural conditions, the leptospira reservoir is primarily mammals from the order of rodents, insectivores, artiodactyls and carnivores. In the natural foci of leptospirosis, voles, hamsters, mice, rats, etc. are of the greatest importance.

Infection of a person with leptospirosis occurs through water infected with sick animals or leptospirosis carriers (bathing, water pistol, performing various work in water). With mass diseases of people in rural areas from the past, this connection with water was so obvious that leptospirosis was called water fever.

Infection of people with leptospirosis is also possible when caring for sick cattle. The causative agents of leptospirosis penetrate the human body through the gastrointestinal tract, damaged skin and mucous membranes.

The clinical manifestations of the disease are very diverse. The disease usually begins suddenly after an incubation period of 5-6 days and is characterized by high body temperature, general weakness, severe head and muscle pains and facial flushing. Jaundice is approximately observed in 10% of patients. In the pathogenesis of leptospirosis, the state of bacteremia, which develops from the first days of the disease, is of great importance. At the end of the first week, leptospira is concentrated in the liver, spleen, lymph nodes, and bone marrow. Subject to the influence of toxic substances resulting from the breakdown of leptospira, parenchymal and fatty degeneration of the liver, focal hemorrhages in the spleen and hemorrhagic nephritis develop.

139

The disease usually ends in recovery.
After the disease is transferred, a stable immunity develops in the body, the mechanism of which is associated with the presence of antibodies. Antibodies appear at the end of 5-6 days of illness.

Laboratory diagnostics involves microscopy of a crushed drop of blood in the dark field of a microscope (Fig. 18).

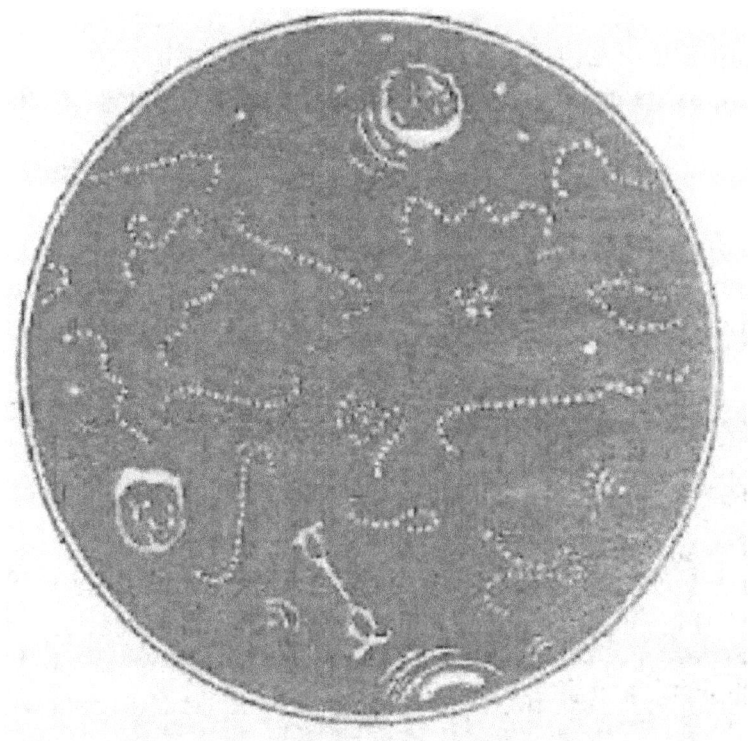

Fig. 18 Leptospira in a dark field

For the treatment, penicillin, tetracycline and antileptospirosis globulin are used.

For prevention in the natural focus of leptospirosis, where human diseases are associated with working in infected water bodies or using infected water for drinking and household needs, it is necessary to prevent people from contacting infected water (prohibition of bathing, use of unboiled water for drinking and domestic needs, etc. .). To eliminate the natural foci of leptospirosis, drainage of swamps and other reclamation measures are recommended.

In urban foci of leptospirosis, deratization and food protection from rats are recommended as preventive measures. Persons working in the outbreak are vaccinated.

Toxoplasmosis is an anthropozoonosis disease caused by Toxoplasma gondii. With toxoplasmosis, the nervous system is affected,

140

lymphatic and endocrine systems, organs of vision. Pregnant women experience abortions or the birth of ugly, non-viable children.

The causative agent of the disease has the shape of a crescent, it is well stained by the Romanovsky-Giemsa method: the cytoplasm is blue, the nucleus is ruby red.

Toxoplasma revealed a sexual and asexual cycle. The sexual cycle takes place in the body of the main hosts - representatives of the feline family, in particular, feline. The causative agent is unstable to environmental factors (high temperature, drying, radiation) and disinfectants. They persist for a long time only in the body of domestic and wild animals, as well as arthropods.

Toxoplasmas are pathogenic for many species of domestic, wild animals and birds. They are found in dogs, cats, pigs, sheep, rabbits, guinea pigs, hares, ground squirrels, rats, mice, monkeys, as well as in some birds. TOxoplasmosis is found in all countries of the world. A person becomes infected from dogs, especially those kept in kennels, from cats, sheep and other animals.

Infection occurs through the digestive and respiratory tract: the pathogen goes down through the mucous membranes (conjunctiva, vagina, mouth); infection is also possible through food and water, are not subjected to sufficient heat treatment and during the bite of arthropods (mites, lice).

Toxoplasmosis can be congenital and acquired. A characteristic sign of congenital toxoplasmosis is hydrocephalus or microcephaly, the presence of foci of calcification in the brain, damage to the organs of vision, cirrhosis of the liver, enlarged spleen, pneumonia, enterocolitis, nephritis, hepatitis. In children who survive, irreversible changes remain in the central nervous system, internal organs, skeleton, which causes deep disturbances in the physical and mental state and can lead to oligophrenia, schizophrenia, epilepsy, idiocy, etc. In the case of asymptomatic toxoplasmosis in the mother, the fetus becomes infected, which leads to abortion and stillbirth.

The most common mechanism of infection is the alimentary route, in which infection occurs when meat, milk, dairy products from animals with toxoplasmosis, raw eggs, sick birds, and water infected with sick animals are consumed.

141

Cases of infection of laboratory workers, obstetricians, gynecologists, surgeons involved in the blood of patients or infectious materials are described.

Adult toxoplasmosis can manifest itself in the idea of a maculopapular rash covering the entire body, with the exception of the palms, soles, and scalp. Quite often, pneumonia, enterocolitis, nephritis, hepatitis, high or subfibrillar temperature are noted. Many patients develop eye lesions.

Pyrimitamine (chloridine) and sulfa drugs are used for treatment.

In the presence of diseases of toxoplasmosis in areas with natural foci, they destroy wild animals, identify patients and carriers among domestic animals, they are isolated and treated.

The meat of sick animals and birds, with the suspicion of the presence of toxoplasma, is subjected to thorough heat treatment, the milk is boiled, the eggs are boiled for 5 minutes. In places where there are cases of toxoplasmosis, systematic deratization is performed, the corpses of domestic and wild animals killed by toxoplasmosis are doused with kerosene or other disinfectants and buried to a depth of at least 1.5 m in special cattle burial grounds.

Among the population carry out health education to clarify the rules of personal hygiene. It is forbidden to eat undigested meat products, especially the liver.

Much attention should be paid to the prevention of infections of a professional nature - workers of the veterinary service, slaughterhouses, meat processing plants, livestock farms, milkmaids, hunters of game wild animals, laboratory workers. Persons of these professions should be periodically examined and, if patients are identified among them, treated.

For the prevention of congenital toxoplasmosis in women who have had spontaneous abortions, premature births, stillbirths, birth of freaks, they are subjected to laboratory examination. People engaged in processing meat and meat products should protect their skin from injuries.

142

3.14. Specific immunoprophylaxis and immunotherapy of infectious diseases

Vaccinal prevention

In the general complex of anti-epidemic measures, they attach great importance to the specific prevention of infectious diseases. The vaccine (lat. Vaca - cow) got its name from the anti-arsenic preparation made from the virus that causes smallpox in cows.

Vaccines are drugs obtained from weakened, killed pathogens or products of their vital activity and used for the active immunization of people and animals for the purpose of specific prophylaxis and therapy of infectious diseases.

Vaccine preparations are divided into the following groups: 1) vaccines from living pathogens with weakened virulence; 2) vaccines from killed corpuscular pathogens (bacteria, rickettsia and viruses); 3) toxoids; 4) vaccines from the products of chemical breakdown of certain bacteria (chemical vaccines).

Live vaccines include vaccines against smallpox, anthrax, rabies, tuberculosis, plague, tularemia, brucellosis, yellow fever, polio, influenza, typhus, measles, etc.

In order to increase the shelf life of the preparations, they are dried in vacuum at low temperature.
Vaccines from killed bacteria include typhoid, paratyphoid, cholera, pertussis, vaccines against Q fever, tick-borne and Japanese encephalitis, influenza, poliomyelitis, leptospirosis, etc.

Anatoxins are made from exotoxins of the corresponding pathogens by treating them with 0.3-0.4% formalin. Widely used are diphtheria, tetanus, staphylococcal and cholera, against botulism. Anatoxins cause the production of antitoxins that neutralize exotoxins, but do not have a detrimental effect on pathogens.

Chemical vaccines are drugs that do not consist of whole bacterial cells, but of chemical complexes obtained by processing the suspension of culture with special methods.

143

The most promising are combined polio vaccines, consisting of different antigens, which allow reproducing antibacterial, antitoxic and antiviral immunity.

Vaccines are injected into the body cutaneously, subcutaneously, intradermally, through the mouth, into the mucous membrane of the nose, pharynx, and as a result, active immunity is created in the body.

Vaccination is carried out taking into account the epidemiological situation and medical contraindications. Contraindications include acute febrile illness, recent infections, chronic infections, heart defects, severe damage to internal organs, pregnancy, and allergic conditions.

Vaccinations provide strong immunity against smallpox, tularemia, yellow fever, polio, diphtheria.

Vaccine therapy. For the treatment of patients with long-lasting infectious diseases (furunculosis, chronic gonorrhea, brucellosis, dysentery, etc.), vaccines from killed microbes, toxoids, extracts from staphylococci obtained by heating are used, they are used with a good therapeutic effect.

In some cases, auto-vaccines prepared from microbes isolated from patients are used.

Serotherapy and seroprophylaxis. Therapeutic and prophylactic serums are released in purified form. Such serums have therapeutic and prophylactic properties, contain the least amount of ballast proteins, have less pronounced toxic and allergic effects.

Serum produced is divided into antitoxic and antimicrobial. The first include anti-diphtheria, anti-botulinum, anti-tetanus, anti-anaerobic infections, anti-snake.

Antimicrobial serums are used against a number of diseases in the form of globulins and immunoglobulins. Immunoglobulins from human blood are used prophylactically against measles, polio, pertussis, viral hepatitis, mumps, chickenpox, scarlet fever. Recently, specific directed immunoglobulins have been produced. They are obtained from serum from a donor immunized against these infections.

144

Such immunoglobulins contain a higher antibody titer.

3.15. Disinfection, disinfestation, disinfestation

There are two main problems of infection in the food industry. The first is food contamination with pathogenic microbes. The second is in the spoilage of products, as a result of which they become unusable. The reason for the low quality of the product and its quick spoilage most often depends on the seeding by extraneous microflora, which enters the products from various sources. Part of the microflora is destroyed by heat treatment of raw materials. The main sources of microorganisms in production are: processed raw materials, water, air, attendants in case of non-observance of personal hygiene rules, premises of the plant and equipment for poor washing and cleaning, as well as insects and rodents.

To prevent the spread of infection, it is recommended to timely remove waste and debris, maintain the cleanliness of rooms and equipment, observe personal hygiene by personnel, and also fight rodents and insects.

However, without the direct destruction of microbes and their vectors, we cannot guarantee prevention in the food industry.

Disinfection is a science that studies the methods and means of destroying the causative agents of infectious diseases at various objects and in various substrates of the external environment. She simultaneously studies the methods of organizing and conducting disinfection measures necessary to ensure their epidemiological effectiveness and quality control of their implementation.

Currently, the doctrine of disinfection combines three sections: 1) the actual disinfection; 2) disinsfestation - the doctrine of the methods and means of combating arthropods; 3) deratization - the doctrine of methods and means of combating rodents.

Along with the concept of disinfection, there is the concept of sterilization.

Its purpose is the complete destruction in any medium or on objects of the most persistent forms of pathogenic and non-pathogenic microbes, including spore-bearing ones. By sterilization, complete provision is achieved, which practically means the absence of signs of life of microorganisms.

145

This is required in surgical, obstetric - gynecological, dental, laboratory, vaccination practice, etc.

Types of disinfection. There are two types - focal and preventive disinfection. Focal - is carried out with the aim of eliminating the focus of infection in the family, hostel, child care institution, on the railway, water transport, in the catering unit. There are two forms of focal disinfection: current and final. Current disinfection is carried out in the outbreak during the entire period of manifestation of the infectious onset. The final is carried out after the cessation of cases of the manifestation of the disease.

Preventive disinfection is carried out in the absence of an identified source of infection, it is carried out as planned. There are three ways to disinfect various items: physical, chemical, biological. Before proceeding directly with disinfection, a mechanical cleaning is necessary. It is especially necessary in the food industry.

During the processing of various food products on the equipment, dense precipitation, scale, and carbon deposits remain. Residues of production fluids. They are a breeding ground for microorganisms. First you need to remove all these deposits using detergents. They must be non-toxic, dissolve well in water, not leave a persistent smell. It is good if they have bactericidal properties.

The sanitary condition of the equipment is estimated by the number of microorganisms remaining on the surface after processing. If not more than 300 cells of microorganisms are found on 100 cm2 of the surface, the sanitary state is considered good, if more than 1000, it is unsatisfactory.

In trade organizations, the presence of insects (cockroaches, flies. Ginger house ants, mosquitoes, rodents - gray and black rats, house mice, field voles) is not allowed.
Thorough disinfection of equipment surfaces is a basic requirement in food processing plants.

Physical methods of struggle. These include the effect of elevated temperatures, radiation. The effect of high temperatures depends on relative humidity conditions. Effective is wet steam sterilization under a pressure of 1 atmosphere.

Influence of infected objects with ultraviolet rays using alpha and beta rays.

146

The rays of the shortest wavelength part of the spectrum have a bactericidal effect, which occurs as a result of photochemical processes that cause cell death. The rays with a wavelength of

253nm act better. They are used to neutralize air and water.

Fire is used to burn infected objects of no value (sputum is sick with tuberculosis, corpses of people and animals who died from especially dangerous diseases). Hot air destroys spores at a temperature of 160-180°C for 1.5 hours. Boiling is often used in disinfection. This is a simple and affordable method of disposal. In this case, all vegetative forms of microbes die as a result of protein denaturation.

In this way, linen, dishes, some food products are disinfected. The layer of water above the laundry should be more than 10 cm with exposure from the start of boiling for 15-20 minutes.

The chemical method of disinfection is the use of chemicals that cause the death of microorganisms on the surface and inside various objects and environmental objects (feces, pus, sputum). Chemicals act more superficially than heat. However, this method of disinfection is used more often. Only those chemicals that have the ability to destroy microbial cells in the environment are suitable for disinfection.

You should know which disinfectant products are used in the food industry. Currently, the following preparations are recommended: "Trias", "Vimol", detergents, "ChM-Z" powder, "Farforin" powder detergents. You can use drugs: sodium hydroxide, sodium carbonate (soda ash), sodium and calcium hypochloride, hydrochloric acid, paraform.

Alkadeks KP 25 - highly alkaline foamy means. It allows you to quickly remove detaturized and native protein, fats, oils. It is used for automatic foam washing of various technological equipment.

Phosphates - have a good rinsing effect. They are active anticorrosive substances. Phosphates are used in almost all detergent compositions. Acid detergents are used to remove hard water scale.

Caustic soda (caustic soda) and preparations containing it have a washing and antimicrobial effect. Use for washing equipment, tanks and bottles.

147

The bulk of yeast and mold dies in 10 minutes when treated with 2% hot solution - 80°C.

Compounds of iodine and chlorine cause the death of microorganisms, including mold spores and bacteria. More often use hypochlorites and bleach. Iodoforms when diluted with water, free iodine is released from the preparations. Sodium hypochlorite is less corrosive to metals than bleach is a powdery preparation with an active chlorine content of 32-38%. Sulfur dioxide is used to disinfect wooden barrels.

Hydrogen peroxide is a strong oxidizing agent that destroys microorganisms and their spores. In the food industry they are mainly used for sterilization of packages and packaging equipment.
Biological disinfection methods are based on the antagonistic action of some microorganisms against other products of their vital activity - antibiotics.

Water disinfection. Water is the basis for the prevention of infectious diseases, but it can become a source of infection. The water should not contain pathogens, microbes and feces of humans and animals. Water disinfection is carried out in order to free it from microorganisms. Physical methods of water purification play a significant role. 1) Filtering through various filters. 2) Boiling for 10-20 minutes. 3) Treatment with ultraviolet rays, which is possible with complete transparency of the water.

Chemical methods of struggle. Ozonation - by adding 0.5 mg O3 per 1 liter, while viruses die in 2 minutes.

Chlorination is the most common, cheapest, and most effective method of disinfecting water. The dosage of chlorine depends on the pH and hardness of the water, the content of organic substances in it. The excess chlorine should be such that after exposure for 30 minutes in the water remained 0.1 - 0.15 mg/l active chlorine. The disadvantage is that the taste of the water changes.

The problem of organizing and combating many infectious diseases transmitted by arthropods has so far stood out in a special large section - disinfestation.

148

Disinfestation is the destruction of harmful insects, which are carriers of various, mainly gastrointestinal diseases (dysentery, typhoid fever, cholera), viruses, and helminth eggs.

In the food industry, flies and cockroaches most often cause harm. Flies are dangerous because they breed in garbage, they sit on sewage, garbage. On the paws and body, they carry a huge number of living microbes (up to 6-7 million), as well as helminth eggs. Every three days, the fly lays up to 600 eggs. That is why the fight against flies must be waged systematically. The main thing is to keep the yard clean, daily waste collection, the correct arrangement of waste bins and the regular treatment of them with a 10% solution of bleach. Containers and waste bins are cleaned at least 1 time per day. In the warm season, they are disinfected.

Disinfectants must meet the following requirements: 1. To have a detrimental effect on insects and in the doses used should not be toxic to humans and warm-blooded animals. 2. To destroy insects with a small dose. 3. To possess resistance to the influence of external conditions (humidity, temperature, everything). 4. Not to change the strength and color of the processed items. 5. To possess a long lasting residual effect. 6. To be cheap.

It is easier to fight flies at the larval stage than when they begin to fly. The destructive method of struggle is aimed at the destruction of arthropods at all stages of their development. Chemical disinfection is carried out only by employees of special institutions.

Deratization - from lat.ratus - rat - a set of measures to combat various rodents that cause significant damage to stored raw materials and finished products. They are carriers of many infectious human diseases: tularemia, paratyphoid, leptospirosis, infectious hepatitis, plague, etc. Preventive and destructive measures are necessary. This device is a rat-proof floor, iron upholstery of the lower parts of the doors to the pantries. Destruction is provided: the installation of traps, traps, the use of poisonous decoys. Employees of special institutions deal with these issues. The use of bacteriological rodent control agents in food enterprises is prohibited.

149

Chapter IV
Human invasive diseases

4.1. Helminthology

Helminthology (Greek Helmis - worm, worm, logos - teaching) - the science of helminths - parasitic worms - and helminthiases - the diseases that they cause.

Helminths are more environmental than systematic, which is very diverse both morphologically and biologically. Helminthology studies all parasitic worms, regardless of who they visit. Therefore, as an applied discipline it can be divided into veterinary, medical and agronomic.

Consider helminthology as a medical science, since they study the pathogenic properties of helminths and develop measures to heal people from diseases caused by parasitic worms.

Helminthiasis, the causative agents of which are parasitic in humans and animals, are classified as anthropozoonoses: they are divided into two groups. The first group is invasions, the causative agents of which develop with the obligatory participation of a person who is an obligate host for helminth. The second group is characterized by the fact that the obligate host of the helminth will be animals, and in humans they are parasitic optional.

Helminths have a certain pathogenicity and to varying degrees cause harm to the body of their host, which manifests itself in the form of a particular disease - helminthiasis.

K.I.Skryabin and R. Schulz (1929), determining the role of helminths, noted that it can be expressed by three factors (triad): mechanical action, toxic effect, as well as inoculation and activation of pathogenic microorganisms. The last factor K.I.Skryabin expressed a very successful aphorism: "Helminths open the gates of infection."

The mechanical effect of helminths is manifested with their fixation, localization in organases and tissues, and migration of larvae throughout the body.

When parasites are fixed to the intestinal mucosa and in various other tissues, they inflict mechanical damage with their armed head organs and cause irritation followed by an inflammatory reaction.

150

The localization of helminths in organs and tissues is the cause of atrophy of the latter.

During the migration of larvae in the walls of the digestive tract, peritoneum, blood and lymphatic systems, in various organs, tissue integrity, hemorrhages, inflammatory processes and other pathological changes occur.

The toxic effect of helminths on the host organism is due to the fact that in the process of life, helminths secrete metabolic products (metabolites) that are poisonous to the body, gland secrets - toxins, as a result of which various pathological phenomena develop in organs and tissues.

Helminthiasis **allergy** is a response of the host organism. According to modern views, this is the most important factor in pathogenesis, and therefore helminthiases are classified as allergic diseases. Helminths, as extraordinary stimuli, are introduced into organs and tissues, as a result of which the body's defenses are mobilized and an active struggle develops between the parasite and the host.

A peculiar form of the allergic reaction in helminthiases is that when a person is infected with helminths with large doses of nematodes, self-passage of helminths occurs with a simultaneous increase in the amount of histamine in the blood.

Allergic reactions develop slowly, mainly on the 5-7th day after infection with helminths. The severity of these reactions depends mainly on the intensity of the invasion (or antigen) and the state of the host organism.

Inoculation and activation of pathogenic microorganisms is manifested in many helminthiases, especially in those cases when pathogens at a certain stage of their development migrate in the host body. There are the following forms of communication between pathogens and helminths in the human body: a) "dormant" or conditionally pathogenic microflora are activated in the host's body; b) when migrating through the intestinal wall, they create a path for microbes located in the digestive tract, and they themselves can enter them into other organs; c) helminths, reducing the resistance of the human body, contribute to the occurrence of infectious diseases; d) parasites also make the outcome of infectious diseases more severe.

151

In addition, in the pathogenesis of helminthiases, there may also be complex variants of helminth-microbial communication with the predominance of any one of the marked elements. It should be borne in mind that the "helminth-microbe" relationship is bilateral in nature, developing during evolution and in some cases taking the form of mutual adaptation, coexistence in any part of the host organism, for example, in the digestive tract, liver and other organs.

The immunity in helminthiases has not yet been studied. The mechanism and general laws of immunity are, in principle, similar to those for infectious diseases, however, immunity tension after relieving remains weak.

The helminths themselves and the life products that they secrete have antigenic properties. Under the influence of helminth antigens, antibodies are produced and the organism sensitizes.

Immunity with helminthiases has specificity, that is, it

manifests itself in relation to those types of helminths, under the influence of which it is formed.

The intensity of immunity depends on the dose of the invasive material (invasion intensity), the type of helminth and its virulence, the individual characteristics of the host organism. Deficient protein nutrition, a lack of vitamins A, B1, B12, B15, C, E and trace elements (selenium, cobalt, copper) play a particularly important role in reducing resistance to helminth infections.

Immunity can be manifested in a decrease in the survival rate of helminths, an increase in the timing of their development, in the limitation of oviposition, and lifespan.

The whole variety of manifestations of immunological conditions in helminthiases is classified according to R.S. Schultz according to the following scheme.

1. Absolute immunity. With this form of immunity, there are no clinical signs of the disease, since the animal is completely immune to helminth infection. Invasive larvae once in the body unable to overcome the intestinal barrier and, passing through it, are thrown into the external environment.

2. Barrier immunity. In this case, the immune animals become infected with helminths, but their larvae linger in the protective barriers of the intestine

152

the wall, skin, liver, lymph nodes, lung tissue, etc. In these barriers, the larvae encyst and die.

3. limiting immunity. In this immunological state, helminths pass through barriers, but the body's defenses quantitatively limit invasion and inhibit the life of helminths.

The described systematization scheme for immunological conditions covers the main most complex biological relationships associated with helminth migration.

4.2. Nematodoses. Nematodoses are diseases caused by helminths from the class of round parasitic worms - nematodes (Nemtoda). This is the largest group of helminthiases from all veterinary helminthology. Nematodes are characterized by an elongated filiform or spindle-shaped body. The length of the body of various species ranges from 1 mm to 10 m. The body is covered outside with a dense layer of cuticle, which forms, together with the muscle tissue lying underneath, the so-called musculocutaneous sac in which the internal organs are located.

The digestive system consists of the esophagus, originating along the entire length of the body. The intestine is a straight tube opening with the anus on the ventral side of the posterior end of the nematode body.

The nervous system consists of a central nerve ring surrounding the esophagus with nerve trunks extending from it.
The female reproductive apparatus is represented by two ovaries, two uterus with oviducts. The male reproductive apparatus consists of the testis and ejaculatory canal.

The biological cycle of the development of nematodes is very diverse. As among all other parasitic worms, among nematodes there are distinguished geohelminths that develop directly, without the participation of intermediate hosts, and biohelminths, the development cycle of which takes place with the obligatory participation of intermediate hosts.

Ascaridosis.
Roundworms are dioecious, females reach 25-40cm, male 15-25cm. Of the most common human diseases caused by helminths, is ascariasis, the causative agent of which Ascaris is found in almost all areas of the globe. Helminth is localized in

the intestine. Here it also lays eggs (up to 24,000 per day), which together with

153

excrement excreted, into water or moist land, where they can remain alive for several years. Under favorable conditions, a larva forms in the egg for 2 weeks, for the further development of which the human body is already required. A person becomes infected by swallowing an ascaris egg with food or water. In the small intestine, the egg shell dissolves and the larva penetrates the bloodstream and the lymphatic system. Once in the liver, the larva then passes through the vein and enters the right atrium, and from there into the lung. Here it is actively introduced into the alveoli, bronchi and rises higher into the trachea, and together with saliva, into the oral cavity, is swallowed again and reaches the intestines. Here it develops to the imaginal stage and again proceeds to the laying of eggs. The number of parasitic individuals can reach up to hundreds.

The only source of invasion is an infected person. Unsanitary conditions and crowding of the population play a huge role in the spread of ascariasis, which causes contamination of soil, water and food by human feces.

Often, infection occurs when unwashed raw vegetables and fruits are eaten. Flies are carriers of eggs. Food workers must keep this in mind.

According to WHO, the prevalence of human ascariasis at one time was: in the Philippines - 85-90%, in Malaysia - 82%, Thailand - 70%, Brazil - 58%, India - 20%, Ethiopia - 58%, Iraq - 98%.
In the Philippines, about 20 million people, with 20 adult roundworms daily lost 2.9g carbohydrates. Consumption of 20 million people corresponds to 1000 fifty-kilogram bags of rice. A child with 50 roundworms releases 10 million eggs daily.

An epidemic of ascariasis in Germany is described when 90% of the population of one city was affected, where 540 worm eggs were found in 100 ml of untreated wastewater. If we talk about the stability of eggs in the external environment, then the larvae in the egg develop in a 5-10% formalin solution, 1% sodium chloride solution.

Being in the small intestine, roundworms creep into the excretory ducts of the liver and pancreas, causing inflammatory processes in them. Sick people suffer from bowel obstruction. Wandering larvae introduce various pathogenic microbes into organs and tissues, causing small hemorrhages. Besides. bronchitis, cough develops.

154

loss of appetite, nausea, vomiting, diarrhea.
In children, teeth grinding, twitching, epileptiform seizures are often observed.

The prevention provides for the protection of the territory from contamination with excrement, rational arrangement and care of latrines and cesspools. The monitoring of water sources. The conducting of health education. At present, gardening plots have a great danger for the spread of ascariasis, where unused feces are used to fertilize the soil.

Enterobiosis

Enterobiosis, a disease of dirty hands, is also widespread. The causative agent of enterobiosis vermicularis or pinworm reaches a small size: female 9-12 mm, male 3-5 mm white. The parasite is localized in the lower part of the small and large intestines of a person. Eggs laid by females mature within 4-6 hours. In the environment, they remain viable for about three weeks. A person becomes infected by swallowing invasive eggs. In the small intestine, hatching of larvae that migrate to the initial sections of the colon occurs. Very often pinworms are found in the blind appendix. Worms reach puberty in 12-14 days. Males usually die after copulation. Females begin to produce eggs. which, however, are not excreted from the genital ducts, but accumulate in the uterus. In this case, parasites gradually descend into the rectum. Oviposition occurs in the folds of the perianal region where the worms exit through the anus. One female secretes from 5 thousand to 17 thousand eggs. After which she dies.

The most common symptom of enterobiosis is acute itching and burning in the anus and perineum, which is caused by irritation caused by parasites during their exit from the rectum to lay eggs. This usually happens at night. In patients, sleep is disturbed, weakness develops and disability decreases. Children whose enterobiosis is especially common lose their appetite and lose weight significantly.
In women and girls, pinworm migration is often complicated by vulvovaginitis resulting from the entry of worms into the vagina and even into the uterus.

For the patient does not pass without a trace and the presence of parasites in the intestine. Sometimes worms can penetrate deep into tissues, penetrating

155

into the abdominal cavity and internal organs. The intestinal phase of enterobiosis is accompanied by acute abdominal pain, diarrhea with abundant secretion of mucus, and sometimes blood.

The source of the spread of the invasive principle into the environment is an infected person. Eggs of parasites accumulate in large numbers on bedding and clothes. They easily disperse along with dust and settle on the surface of a wide variety of objects, including food. When combing itchy spots, many eggs remain on the hands and under the nails. This leads to the widespread occurrence of autoinvasion during enteroyosis, as a result of which the disease sometimes lasts for a long time. Enterobiosis is especially widespread in children. A decisive role in the fight against enterobiasis belongs to preventive measures based on strict observance of sanitary and hygienic rules. In these cases, it is possible to eliminate the disease even without the use of drugs.

In order to prevent, thoroughly damp cleaning should be periodically carried out. Then boil brushes and rags. Periodically, an examination for enterobiosis of all personnel of food industry workers should be carried out. Especially those who come in contact with products, ready meals.

Trichocephalosis

Widespread nematodosis caused by the species Trichocephalus trichiurus. The length of the parasites can reach 4-5cm. Whipworms are localized in the cecum, and with severe invasions and in the adjacent sections of the digestive tract. Egg development is carried out in the external environment. Eggs remain viable and invasive for up to 1.5 years.

Human infection occurs by swallowing invasive eggs. Hatched larvae penetrate the villi of the intestinal mucosa, where they are up to 10 days. After this, the larvae again enter the intestinal lumen and reach the cecum. In the human intestines, they live up to 5 years.

Severe infection often entails serious disruption of the digestive tract, accompanied by nausea, pain, and a decrease in acidity. Violation of the integrity of the mucosa opens the gate to secondary bacterial infections. Often with trichocephalosis observed

156

reaction from the nervous system: headaches, dizziness, and sometimes seizures.

In the epidemiology of trichocephalosis, a crucial role is played by an infected person, who is the only source of the spread of the invasive principle in the external environment. Whipworm eggs scattered in the soil easily fall on vegetables and fruits. Possible contamination with contaminated hands. Trichocephaliasis is found almost throughout the globe, with the exception of the Far North and arid regions.

Trichinosis

Trichinosis - an acute or chronic anthropozoonous invasive disease with pronounced allergic effects caused by nematodes of the family Trichinellida, type Trichinella spiralis.

Adult trichinella parasitize in the small intestine of animals and humans, and larvae in the striated muscles of the same organisms. To date, more than 100 species of mammals that are hosts of Trichinella have been registered.

The most common trichinosis occurs in pigs, dogs, wolves, foxes, cats, bears, rats, mice. Helminthiasis is registered in marine mammals - whales, walruses. seals. The economic damage from trichinosis is very large: trichinosis in humans is very difficult, poorly treatable and often fatal. Pathogen - the female reaches a length of 3-4mm, viviparous.

In trichinosis, the same animal is first definitive, and then an intermediate host of helminth. Animals and humans become infected by eating trichinosis meat, which contains live encapsulated trichinella larvae.

These larvae develop rapidly in the small intestine until puberty. Trichinella females penetrate their mucous membrane with their head end and, on the 4th day after infection, hatch live larvae. One female hatching up to 2100 larvae. The larvae penetrate the lymphatic, then into the circulatory system and are carried throughout the body by a current of blood. They linger in the striated muscles, penetrate the muscle fibers, grows, then coils. In muscles, the encapsulated trichinella larvae can remain viable for 25 years (Fig. 19).

157

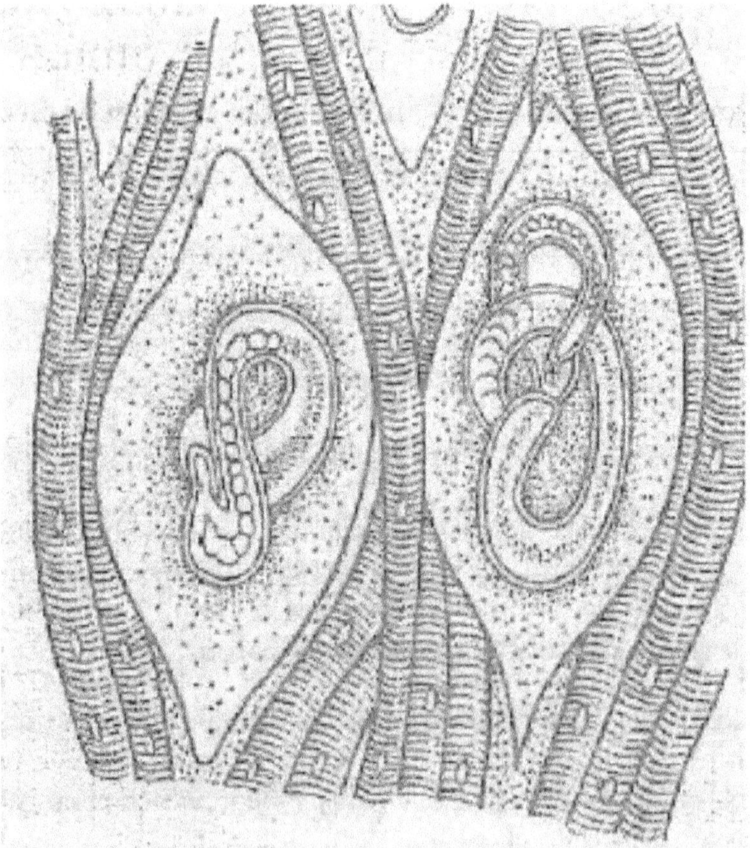

Fig. 19. Trichinella spiralis

Helminthiasis is ubiquitous. Of great practical and health importance is pig trichinosis. Pigs become infected by eating corpses of infested rats. Infection of pigs often occurs with caudophagy, when pigs eat the tails of infected pigs. For this reason, trichinosis from natural foci passes into synanthropic.

Trichinella larvae injure tissues and cause hemorrhages during the period of migration. The waste products and decay of larvae, as well as the products of destroyed tissues cause intoxication of the body.

Coincidence of human trichinosis is usually preceded by an incubation period, the duration of which varies from several days (5-8) to 3-4 weeks. The basis of the mechanism of the pathogenesis of trichinella is an allergic reaction of the body caused by sensitization by the products of metabolism and decay of dying parasites.

One of the most characteristic symptoms of the disease is swelling. The first to appear is edema of the eyelids, which in severe cases of the disease spreads throughout the face, to the neck, trunk and even limbs. At the same time, a fever develops, which reaches a maximum in 2-3 weeks. An equally constant sign of the trichinosis is muscle pain.

158

They occur, as a rule, in the ocular, chewing and cervical muscles. Quite often, the soreness is felt in the shoulder, lumbar and calf muscles. A very severe pain during movement can lead to muscle contracture.

Sometimes the trichinosis is accompanied by an upset gastrointestinal tract: diarrhea, severe abdominal pain, and vomiting.

The trichinosis refers to the diseases with natural foci. Currently, there are two types of foci: natural and synanthropic. Moreover, synanthropic foci are constantly replenished due to the influx of invasive origin from natural foci.

Trichinosis is found almost everywhere. The only exception is Australia, where there are isolated cases of the disease throughout the territory.

A decisive role in the fight against trichinosis belongs to preventive measures. These include: mandatory microscopic examination of meat of domestic and wild animals eaten; culling and destruction or technical disposal of infected carcasses; the creation of sanitary conditions that exclude the possibility of infection of livestock and other domestic animals; the destruction of rats and mice, which are an important component of synanthropic foci of the trichinosis.
Strict deratization measures at pig farms, slaughterhouses, meat processing plants prevent the spread of the trichinosis.

4.3. Tapeworms –Cestoda.

The tapeworm class unites about 3,500 species of parasitic flatworms, living mainly in vertebrates. A characteristic feature of tapeworms is their lack of a digestive system. Their body is flattened and elongated. In most cases, it is not divided into numerous segments - proglottids. All adult worms in the adult stage live in the digestive system of animal hosts.

Cestodes were first described as an independent class of flatworms as early as Rudolfi in 1819. An important role in the development of cestodology in our country was played by the publication of the multi-volume edition of the Fundamentals of Cestodology, carried out under the guidance of Academician K.I.Skryabin.

159

The morphological diversity of tapeworms is very large, although it does not in any way affect a single plan of the structure characteristic of the whole group as a whole.

In most cases, the cestode body is divided into three segments: the head — scolex, the neck, and the articulated strobila (Fig. 20).

Scolex in the form of a compact formation has a variety of forms, size, structure. Scolex chains are usually more or less rounded, with 2-4 suction cups with muscle walls, which can be armed with hooks. At the apex, the scolex is often equipped with a special muscular outgrowth — the proboscis, which carries weapons in the form of one or more rows of hooks.

Head with suction cups

Behind the scolex is an unsegmented area of the body - the neck - the growth zone in which the formation of segments occurs. As new segments form between the cervix and the anterior segment, older segments gradually move to the posterior region, and the youngest segment is located at the cervix.

Mature proglottid with the uterus filled with eggs

Reproductiove organs in the proglottid

Neck

Suction cup

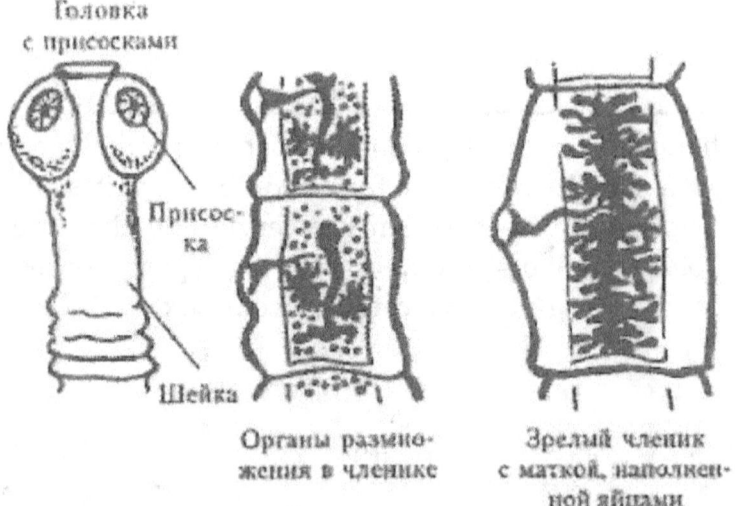

Figure 20. Body structure of tapeworms.

The linear dimensions of cestodes vary within very wide limits. Among them there are giants when the worms reach 25 m. However, representatives of this class of 8 mm in size are also known. In many cestids, the pneumatic-type attachment organs are combined with various types of chitinoid hooks.

160

An increase in the number of segments and elongation of the strobila in most species is carried out according to the principle of intercalary growth. New proglottids will form in the bud zone and shift previously formed segments back. The proglottids that make up the posterior portion of the strobila are constantly torn away from it. Separated segments are often capable of active movement.

The body of the cestode covers the skin-muscle layer, consisting of a cuticle and a basement membrane. The nervous system of cystodes consists of several nerve nodes located in the scolex. Tape parasites do not have a morphologically expressed digestive system, and they feed by absorbing food throughout the surface of the body.

The male and female genital openings usually open side by side. The middle part of the strobila is made up of segments with a well-developed functioning reproductive system. These are the so-called hermaphroditic joints. At the posterior end of the strobilus there are "overripe" segments. It is these proglottids filled with eggs that break away from the body of the worm and are carried out.

Very dangerous human cestodoses are the diseases caused by larval stages (armed cestodes). Among them, echinococcosis and alveococcosis are distinguished by the breadth of distribution and exceptional pathogenicity.

Echinococcosis - anthropozoono proceeding is usually asymptomatic in the body of the intermediate host. The causative agent of the Echinococcus granulosus. Cattle, pigs, deer, less often horses get ill with it. The echinococcosis also affects people. The disease is common almost everywhere. Especially often it is registered with us in the southern regions of the country. The final host of the pathogen is the dog and the members of the Canidae family. The tape stage of echinococcus parasitizes in the small intestine of dogs, wolves, jackals and less often foxes. The echinococcus in the tape stage is a very shallow cestode, the strobila of which has a length of up to 6 mm and consists of 3-4 segments.

161

In this case, the mature segment exceeds the rest of the strobila in length.

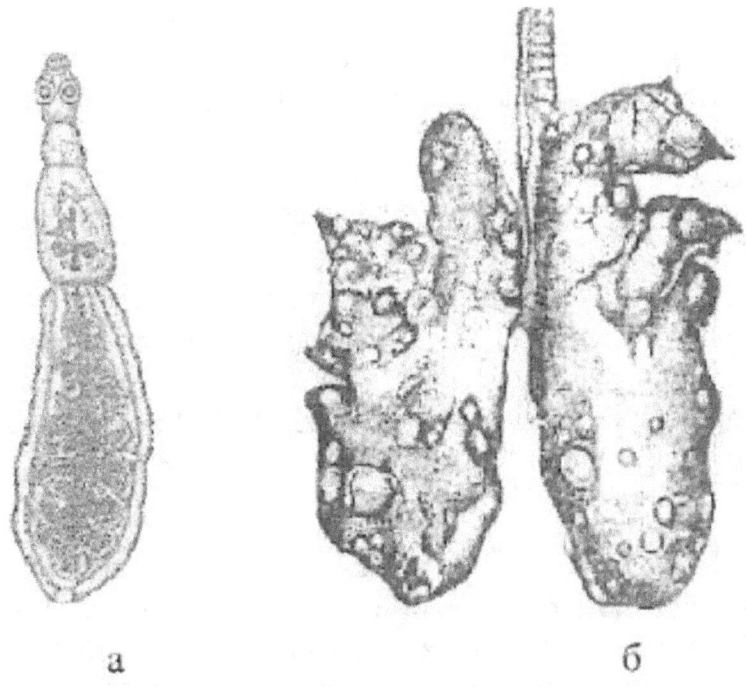

a   б

Fig.21.a –Echinococcus granulosus (sexually
mature form); b – mild, larval stage affected

The scolex is equipped with a proboscis armed with 26-40
hooks. The number of eggs in the mature segment averages
about 800, and sometimes reaches several thousand. The
segment, torn off in the intestines of the final host, filled with
parasite eggs, autonomously moves along the substrate, rises
along the plant steels and scatters a huge number of eggs (Fig.
21).

Recently it was established that E.granulosus adapted in
different geographical zones to parasitism in a certain circle of
hosts. In the west of the European part of the country, a swine
strain of the pathogen is widespread, which actively functions
in the host circle: dog - pig - dog.
In the southeast of the country a sheep strain is spread: dog -
sheep - dog.

The circle of intermediate owners E.Granulosus very wide. These are almost all domestic and wild ungulates, marsupials and rodents.

A human becomes infected by eating fruits, vegetables and any other food products that have been affected by the eggs of this parasite.

In the digestive tract of the intermediate host (human), the oncospheres are freed from the egg shells and invade the intestinal mucosa. Most often, they enter the liver, where the development of the larvocyst (bladder) begins. The parasites can be localized in other organs, but this is observed much less frequently.

162

The second most frequent lesion after the liver is lungs. The bubble reaches very large sizes up to 15 cm across. The mother bubble contains a large number of daughter bubbles.

A human in the life cycle of echinococcus serves as an intermediate host. In the early stages of the development of the parasite, the disease is asymptomatic. The appearance of the first clinical signs, the nature of which is largely determined by the localization of the bladder, is often separated from the moment of infection with a period of several months, or even years.

The pathogenesis of the echinococcus is determined by the sensitization of the host organism and the mechanical action of the parasite on surrounding tissues. An allergic reaction in mild cases manifests itself in the form of skin itching, urticaria. With complications associated with rupture of the bladder, when a large mass of fluid enters the host, shock occurs. Especially dangerous is the destruction of the dead and festering larvocists. Usually, purulent peritonitis is fatal for the patient. With any violation of the integrity of the maternal bladder, there is always the most serious danger of a large number of new bubbles.

In case of liver damage, gradually increasing pains are observed, the volume of the affected organ increases, signs of exhaustion, anemia, icteric staining of the integuments, ascites appear. The development of larvocysts in the lungs is associated with cough with sputum, a slight increase in temperature, severe shortness of breath. As the symptoms intensify, the clinical picture more and more resembles tuberculosis. Rupture of the pulmonary bladder can have serious consequences. If the fluid enters the pleural cavity, then the patient develops shock, accompanied by severe anaphylactic phenomena. In this case, death often occurs.

In the spread of echinococcosis, a decisive role belongs to definitive hosts - dogs, and somewhat less often to wolves, shakash and other predatory and intermediate hosts - large and small cattle, pigs. With close contact of a person with invasive animals, the probability of infection becomes very high.

Measures to prevent infection of people with echinococcosis are based on strict observance of the rules of personal and public hygiene. When slaughtering the livestock, destruction or disposal of organs affected by larvocysts is important.

163

Alveococcosis laurel. A natural focal disease of rodents, as well as humans, caused by cestode larvae Alveococcus multilocularis, localized in the liver and rarely in other organs. In the tape stage, alveococcus parasitizes in the small intestine of the fox, foxes, wolves and dogs.

The definitive hosts (arctic foxes, foxes, wolves and dogs), infected by A.Multilocularis, scatter into the external environment, along with feces, mature segments of the parasite filled with eggs.

As a rule, human infection occurs in natural foci and is most often found in hunters. Often there is an invasion in persons associated with the processing of animal skins. The source of the disease in humans can be sledding, hunting and shepherd dogs. A serious danger is the use of water from small, stagnant forest ponds, wild berries, and edible plants collected in areas inhabited by wild predators.

In the intermediate host, alveococci develop very quickly and intensively. In the experiment on day 41, the rat liver is a continuous alveolar formation with negligible remnants of liver tissue.

The final hosts become infected by eating rodents invaded by larval alveococcus. In the organism of the definitive host, the parasite fully develops in 30-33 days, and the life span of the cestode usually does not exceed 3 - 3.5 months.

For alveococci, the characteristic is infiltrating growth. They are more pathogenic than echinococcus. Parasites compress tissues and cause atrophy of affected organs.

For the purpose of preventive care, special attention should be paid to the prevention of infection of people in the natural foci of alveococcus. Skinning from foxes and arctic foxes should be done with great care.

Cestodoses in which a human is the ultimate host of the pathogen

Three anthropozoonoses belong to this group of teniidoses: bovine cysticercosis (cattle finnosis), cellulose cysticercosis (pig finosis), and diphyllobothriosis (wide tape). These helminthiases are caused in

164

These helminthiases in animals are caused by the larval stages of the tenoid, while sexually mature cestodes parasitize in humans.

The causative agents of teniidosis are both tape, imaginal, and larval, larval, stages of the tenoid.

Cysticercoses cause significant economic damage due to the rejection of heavily affected carcasses, a decrease in the quality of invaded meat and the cost of rendering it harmless. The health value of tenidiasis is great. Unarmed tapeworms and armed tapeworms, parasitizing in the human intestines, significantly reduce its working capacity, slow down its development. Besides, the cysticerci of the unarmed tapeworms parasitizing in the brain and eyes of a human, they can cause significant visual impairment and mental activity, and sometimes cause death. Each of these helminthiases has its own distinctive features, although the measures to combat them are identical.

Phinoses are chronically occurring human anthropozoon diseases. Tapeworms vary in length from 1.5 m to 9.0 m. It should be noted that they are introduced into the body of the tol by the alimentary route. The habitat of sexually mature cestodes is the intestine. The fertility is very high - up to 400 million eggs per year. Life expectancy is up to 35 years. For 35 years, a person lives with a helminth. Thus, the cestodes at all stages of development are parasites.

By nature, it was destined for a person to become the ultimate host, while the only one that spreads these diseases. Food industry workers need to know these diseases, because without knowing the nature of the problem, they can cause them to spread.

Cysticercosis in the bovine cattle is caused by the larval stage of the Taeniarhyncus saginatus cestode parasitizing in the human intestines. Cysticerci (armed tapeworms) are localized in the intermuscular connective tissue of skeletal muscles, heart, tongue, less often in the tissues of the parenchymal organs of cattle.

Mature stage of T.saginatus reaches 10m or more. Scolex is unarmed, there are suckers on the proboscis, with the help of which the parasus is attached to the intestinal wall. The number of mature progloditis excreted in the feces per day is 6-8 segments. In each segment, the number of eggs reaches 145-175 thousand. As a result, from 175 thousand to 4 million 900 thousand eggs are allocated per day for one year.

165

The intermediate owner is cattle, buffalo, yak, zebu, and, according to the latest data, gray deer (Fig. 22).

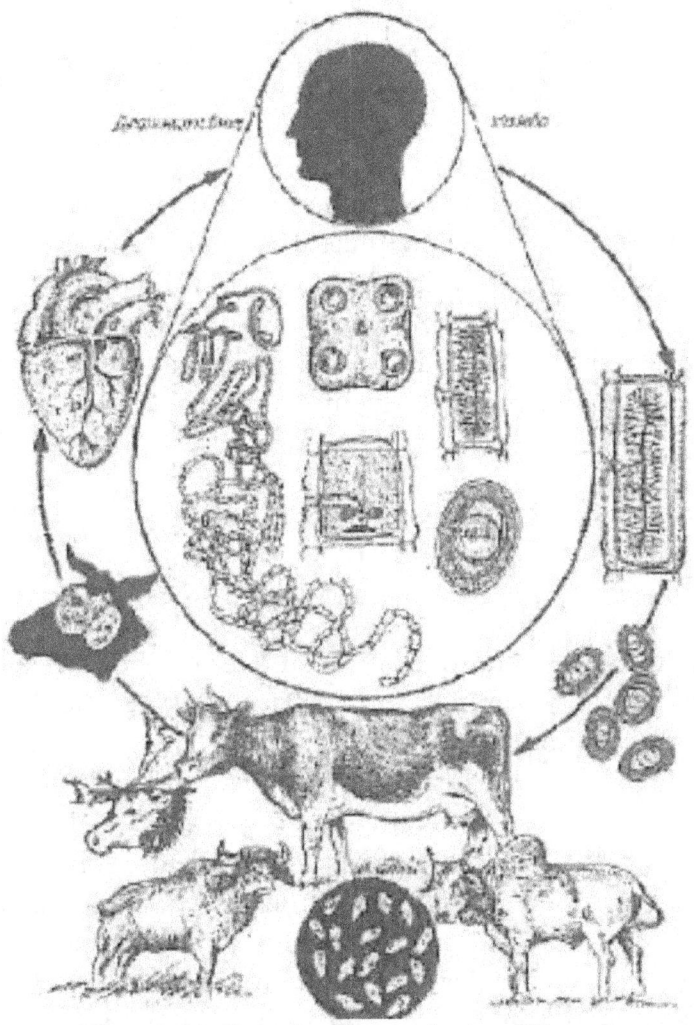

Figure 22: Development chart of T.saginatus

From the intestines of a human invaded by Taeniarhynchus saginatus the mature segments secrete into the external environment passively with excrement, or actively crawling out of the anus. Infection of the intermediate host occurs when it swallows the oncospheres (eggs). In the intestines of cattle, the embryo leaves the egg and penetrates into the intestinal capillaries. And in the future, by the hematogenous route, it can be entered into any organs where an invasive cysticercus forms after 4.5 months. Cysticerci develop primarily in the intermuscular connective tissue. In some cases, they achieve invasiveness in the subcutaneous tissue, adipose tissue, liver, lungs, heart.

A human becomes infected with teniarichnosis when eating dishes from the raw or insufficiently cooked meat (kebab, basturma, steak, etc.). When cysticerci enter the human digestive tract, they undergo gastric juice and bile inversion of the scolex, which is attached to the wall of the upper chatsi of the small intestine using suction cups.

166

From the moment the cysticerci enter the human intestine to the formation of a sexually mature cestode, an average of 3 months passes.

The teniarinhoz is found in almost all countries of the world. Especially often the population of Africa, Australia and some parts of Asia suffers from the teniarinhosis. The foci of disease are available in the Caucasus.

The fight against teniarinhozm is a set of measures aimed at improving the terrain (neutralization of wastewater, destruction of eggs on the soil surface, strict observance of sanitary norms and rules). Periodic mass examinations of the population in the foci of the disease and treatment of identified patients, the strictest veterinary control over the implementation of the rules for slaughter of animals and the disposal of measled meat. Food industry workers are strictly forbidden to purchase meat on the side, without veterinary examination.

Porcine cysticercosis is caused by the larval stage of the cestode Taenia solium parasitizing in the human intestines. The cysticerci themselves are localized in various organs and tissues. Most often they are found in pigs in the muscles, heart, brain, eyes, liver and lungs. In humans, in the brain and eyes.

The porcine cysticercosis is recorded in the central regions of the nonchernozem zone. Intermediate hosts are domestic pig, wild boar, bear, camel, dog, cat rabbit, hare, as well as humans.

A human is the only definitive host that periodically excrets mature segments with feces. Due to the life span of the parasite, one patient can infect a large area with tapeworm eggs. This contributes to the lack of comfortable toilets, their unsanitary condition. In the external environment, the joints make active movements, while the eggs are expelled from the uterus through the destroyed edge of the proglottid. A person becomes infected by swallowing the eggs of T.solium with food or water (Fig. 23).

Thus, the final development of the parasite occurs in the human body, which becomes infected by ingestion of the formed cysticerci, which are in unboiled or unroasted meat.

In the human gastrointestinal tract, the bladder membranes are digested, and the scolex is inverted and attached to the intestinal mucosa, penetrating into it with its own hooks.

167

The formation of the strobile begins, and after 2-3 months, mature joints appear in the pork tapeworm. The life span of a parasite in the human body is estimated for years. The longevity of cysticerci in pigs is 3-6 years, after which they are wrinkled, impregnated with lime and die.

The tape stage does not exceed 3 m in length. The Scolex is armed with a double crown of hooks, the number of which ranges from 22 to 32.

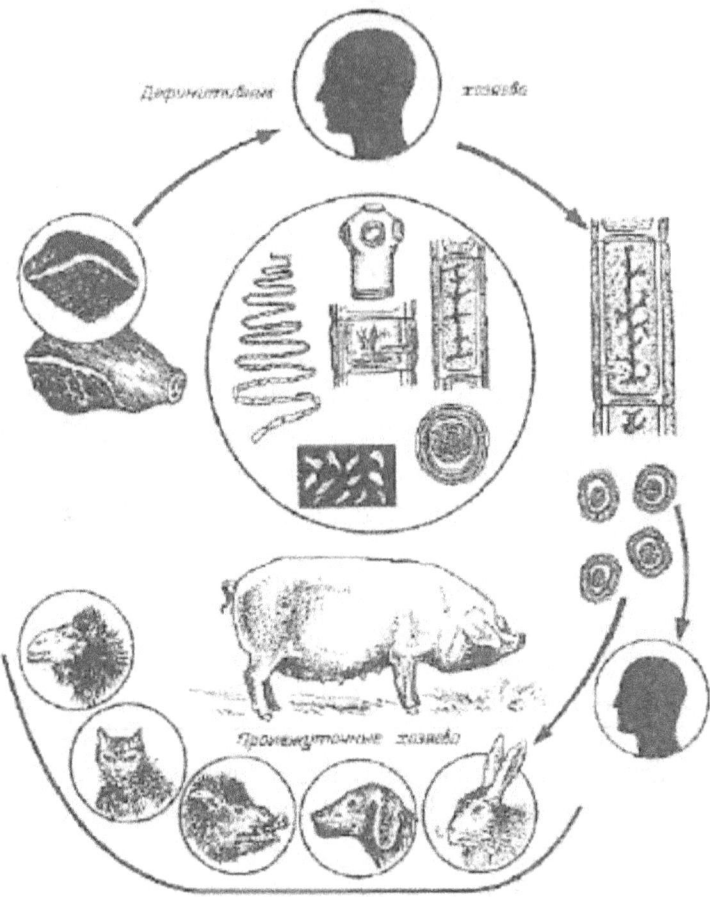

Figure 23: Development Diagram of T.solium

Oncospheres in the intestinal lumen penetrate into the mucous membrane and are carried through the body with a blood stream. The parasites are localized in almost all tissues and organs, however, preferring intermuscular connective tissue. In humans, they are often located in the eyes and central nervous system. The cysticercus formation ends 60-70 days after infection.

The external clinical manifestations of the disease are more pronounced than with teniarinhoz. The patient has an appetite disorder, nausea, stool, abdominal pain, dizziness, headaches, and sleep disturbance. However, the main danger of teniosis is that a person

168

acts in this case not only as a definitive host, but also as an intermediate host for T.Solium. the infection of a human by the armed tapeworm, dubbed cysticercosis, can be carried out in two ways. Firstly, by ingestion of eggs scattered in the external environment. Secondly, very often there is an autoinvasion, the consequences of which can be extremely severe. The segments rejected from the strobila contain a huge number of completely invasive eggs. However, with vomiting, which is very common with teniosis, as a result of intense intestinal antiperistalsis, such areas of the strobila can enter the stomach. Here, under the influence of gastric juice, a massive hatching of the oncospheres occurs, which are subsequently carried throughout the body, causing various lesions. The development of larvae in the subcutaneous tissue and skeletal muscle, as a rule, is asymptomatic or is accompanied by unpleasant pain, which is often misdiagnosed as rheumatism.

Larvae are dangerous in the brain and spinal cord and eyes, but it is in these organs that they are localized most often. Intracranial pressure rises. Sometimes there is a blackout of consciousness, mild mental disorders. The disease is often fatal. The intraocular cysticercosis, which causes deterioration, and sometimes complete loss of vision, in the absence of treatment leads to atrophy of the eye.

The spread of the invasive principle in the external environment is carried out only by an infected person. Infection of a person with larval stages, as noted above, occurs when meat products are consumed. This is most often observed in rural areas with non-compliance with sanitary standards. Therefore, the health education should be systematically carried out among the general population by both veterinary and medical personnel. In order to engage the population in the active fight against helminth infections, all the opportunities that are provided in specific conditions are used (lectures, talks, press, radio and television, demonstration of popular science films, publication of popular science literature). All these measures must first be carried out in areas unsuccessful for tenyidosis.

Diphyllobothriasis refers to natural focal diseases, the spread of which is associated with certain environmental conditions.

169

It is cased by the appearance of a lentec Diphyllobothrium latum. Humans, dogs, cats, foxes, polar foxes and martens (definitive owners) get ill. The length of the strobile averages about 10m. In fur animals, its size usually does not exceed 1.5 m. His body consists of 3000-4000 individual segments. The development scheme of the parasite is shown in Fig. 24. Pollution of lakes and rivers with invasive material - feces of sick people - comes from latrines built above water and along the banks of water bodies, as well as from steamers and barges and other vessels; fecal particles containing lentets eggs are carried into the water bodies from the soil by rain and melt water. The population living on the banks of lakes more often infected with diffilobotriosis than living on steep river banks.

The life span of this parasite can be very significant - 20 years or more. Usually, in the intestines of a human one or, in extreme cases, several individuals live. The intermediate hosts are the bipedal cyclops, and additionally, freshwater fish (pike, perch, ruff, burbot, trout, etc.). The eggs are secreted into the external environment along with the stool of definitive hosts and must be released into the water for further development. After 3-5 weeks, larvae emerge. In the future, the larva is swallowed by intermediate hosts, in the body of which development proceeds.

But for his further development, it needs to get into the body of an additional host - fish. In the intestines of the fish, the crustacean is digested, and the parasite penetrates through the intestinal wall into its abdominal cavity, muscles, subcutaneous tissue, where it forms in the next stage, reaching 6 mm in length,

in the body of which the pathogen reaches the invasive stage. If this fish is eaten raw or half-baked (dried) by humans, then the larva, freed from the muscles of the fish, attaches to the wall of the small intestines. And here, 40 days later, it develops into an adult broad ribbon, the formation of segments begins and their release into the external environment. Thus, a human becomes the definitive host.

S.P. Botkin was the first to reveal a connection between parasitization of a broad ribbon and the development of anemia, with the manifestation of tachycardia in 1884.
In Finland, there has been an increase in the number of patients with diphyllobothriasic anemia during the Second World War.

170

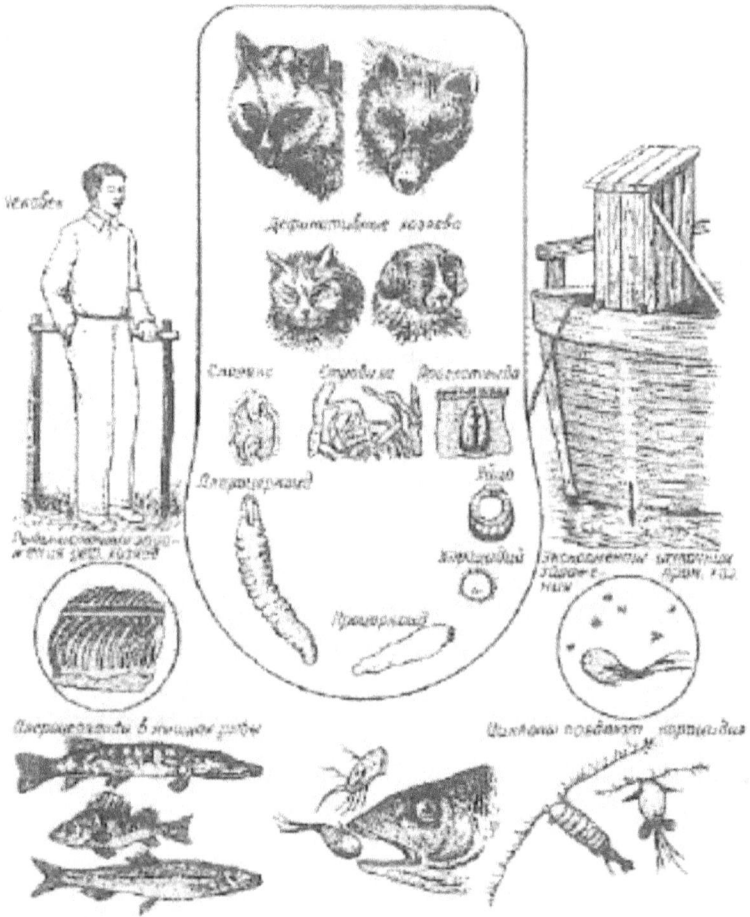

Fig. 24 Life cycle of Diphyllobothrium latum

Being localized in the human body, the parasite causes only weak intestinal disorders. However, in many cases a very severe course of the disease is also observed. It is caused by the mechanical action of cestodes on the intestinal mucosa and sensitization of the infected person's body by their metabolic products. In addition, the products of lentic exchange cause changes in the microflora of the intestines of the host, which leads to an almost complete cessation of the folic acid biosynthesis by intestinal bacteria and a decrease in the amount of vitamin C. Severe anemia develops. The patient develops weakness, severe dizziness, drowsiness. Often develop pathological changes in the integument, swelling. The liver swelling and enlarged spleen is observed. In children, seizures, convulsions are observed.

The fight against diphyllobothriasis should include a number of activities. First of all, it is necessary to create conditions that prevent the infection of water bodies with uncontaminated wastewater from settlements and ships that run along rivers and lakes. Individual prevention consists in eating only well-prepared fish.

171

Food industry workers should be extremely attentive to the reception and sale of fish.

4.4.Trematodoses.

Trematodoses are parasitic worms related to the type of flat flukes. All trematodes are parasites localized in various organs and tissues of animals and humans. Trematodes are biohelminths; the first intermediate hosts are mollusks, both aquatic and terrestrial (land).

A large number of species of trematodes are known that cause a dangerous disease in humans, domestic and commercial animals. Settling in a variety of vertebral organs, trematodes exert a variety of pathogenic effects on the host.

So, serious violations arise as a result of the migration of trematodes through the host tissues, during their nutrition, attachment, etc. Parasites can disrupt the integrity of tissues, cause obstruction of the ducts (bile ducts of the liver, ureters, etc.) and blood vessels, and may disrupt the secretory function of the intestinal epithelium. No less important is the toxic effect, which leads to the poisoning of the host with parasite metabolism products that are poisonous to it, which enter directly into the bloodstream, into the thickness of tissues or into the lumen of internal organs. Finally, the trematodes can provoke the appearance and growth in the host organism of various kinds of neoplasms, including malignant ones.

Here we will restrict ourselves to considering only one trematodose, which is of the greatest medical importance.

Opisthorchiasis refers to natural focal diseases, since the pathogen is pierced in nature without human intervention.

Two large natural foci of opisthorchiasis are known in Russia - in Western Siberia and the Perm Region. In addition, group and single cases of infection of people were found in the area of the coast of the Black and Caspian Seas, Lake Ladoga, Volga basin, Don, Donets, Northern Dvina, Neman. The disease is also registered in Europe, i.e. in Austria, Hungary, Holland, France.

The causative agent of opisthorchiasis - catliver fluke – Opisthorchis felineus. The parasite is localized in the bile ducts of the liver and gall bladder, and sometimes in

172

ducts of the human pancreas and many mammals (cat, dog, pig, some fur-bearing animals). This is a small fluke, the body of which does not exceed 8-18 mm in length.

O.felineus was discovered by the German parasitologist G. Vogel. The development of the egg is entirely carried out while moving along the loops of the uterus of the worm and eggs that already contain the formed larvae are taken out of the intestine of the worm. To carry out the life cycle, eggs must fall into the water (Fig. 25).

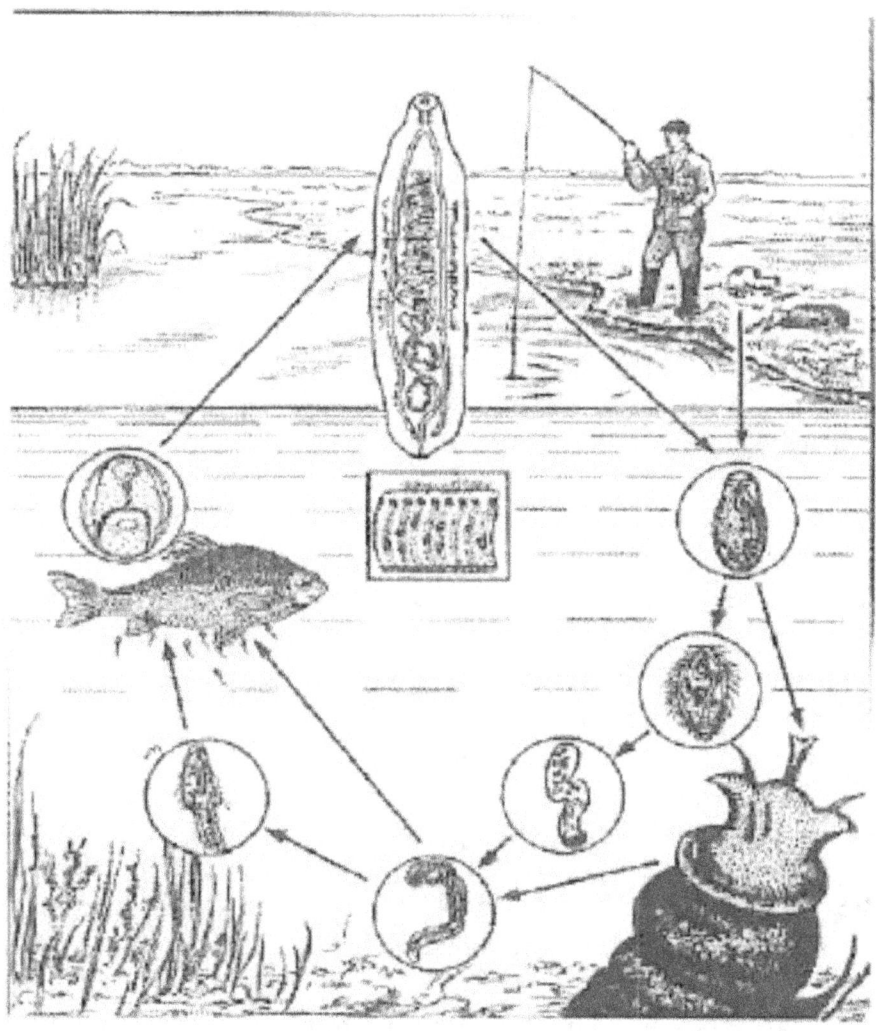

Fig. 25. Development chart of Opisthorchis fileneus

They are devoured by freshwater mollusks, in the intestines of which hatching of myrocidia occurs, which develops to circaria and leave the mollusk's body. Cercariae are very mobile and are good swimmers. They are characterized by negative phototaxis and geotaxis with the same sign. This allows them to stay in that very zone of the reservoir in which the meeting with the second intermediate hozain is most likely, the role of which is played by many species of cyprinids - roach, tench, ide, bream, rudd, etc.

after 6 weeks they become invasive (metacercariae).

173

A definitive host, including a person, becomes infected with opisthorchiasis by eating raw, frozen or dried fish invaded by metacercariae. In the intestines of the definitive host, metacercaria are released from the cysts and migrate through the bile duct to the liver, gall bladder and pancreas. 3-4 weeks after infection, flukes reach maturity and begin to secrete eggs. Their life expectancy is several years.

A opisthorchiasis usually proceeds as a chronic disease with exacerbations and remissions. In the "acute period" there are severe pains in the pit of the stomach and in the right hypochondrium, loss of appetite, emaciation, dizziness, headaches. The liver is often enlarged, compacted, the gall bladder is enlarged. There are violations in the functioning of the pancreas.

It is noted that a year after moving to the focus of opisthorchiasis, the infection is 11.5 to 17.9% of people. After 1.5 years - 42%, after 5 years - 46.5% of people. In Russia, there are more than 2 million people affected by opisthorchiasis. The invasion of the population of Western Siberia is 51-82%. In some areas - 95%. That is why opisthorchiasis is becoming a major environmental health problem.

Food industry workers must feel responsible for the fact that the prevention of these diseases largely depends on how seriously and responsibly they take their responsibilities. Accurate compliance with the requirements of sanitation and hygiene, receipt of goods according to veterinary documentation will ensure the exclusion of infection by the above diseases. Of great importance is compliance with the requirements of food preparation technology.

Chapter V
Sanitary protection of the territory of the country from the importation and spread of infectious diseases

Sanitary protection of the country's territory is a system of national measures aimed at preventing the importation of quarantine and other especially dangerous infectious diseases into the country's territory from other countries, localization and elimination of foci of these diseases in the event of their occurrence in the Russian Federation, including in endemic natural foci, as well as the prevention of the import and distribution of goods potentially hazardous to public health.

174

More than 350,000 diseases of cholera, 2500 cases of plague and 500 cases of yellow fever are registered annually in the world. In 1994-1996, large outbreaks of plague were noted in the immediate vicinity of the borders of Russia with China and Mongolia, as well as in India and Madagascar. Over the past 6 years, 2.4 million people fell ill with cholera in the world, of which more than 60 thousand died.

In the Russian Federation there are 12 natural foci of plague with a total area of over 30 million hectares. An increased risk of contracting this disease is more than 20 thousand people. The most active foci are located in the Astrakhan region, Kabardino-Balkarian, Karachay-Cherkess, Chechen Republic, the Republic of Dagestan, Kalmykia, Altai. In 1992-1996, 1075 strains of the plague pathogen were isolated from rodents - carriers and carriers of infection, and the area of epizootics was more than doubled.

The largest outbreak of cholera during the period of 7 pandemics resulting from the importation of infection by pilgrims was recorded in the Republic of Dagestan in 1994. During the outbreak, 1,119 people fell ill and more than 1,200 carriers were detected. The total cost of conducting only diagnostic, treatment and anti-epidemic measures amounted to more than 2 billion rubles, excluding damage to the economy of the republic by conducting restrictive and other measures.

In the last two years, 15 cases of cholera infection, both imported and local, have been recorded (Moscow, Astrakhan, Moscow, Rostov regions, the Republic of Dagestan and the Chechen Republic). Already in April 1997, the import of cholera from the countries of the Indian buscontinent was registered in Moscow.

Sanitary protection of the country's borders from the importation of infectious diseases is one of the widely known and long-used areas of preventive work.

Sanitary protection of the territory is an integral part of the system for ensuring the sanitary and epidemiological well-being of the population of the Russian Federation, consisting of a set of organizational, sanitary-hygienic, anti-epidemic, therapeutic, economic, technical and other measures that ensure the prevention of the importation and spread of quarantine diseases, contagious viral hemorrhagic fevers , malaria and other infectious diseases transmitted by mosquitoes dangerous to humans,

175

localization and elimination of cases of these infections when they are imported or detected in the territory of the Russian Federation, including in endemic natural foci, as well as the prevention of the import and distribution of goods that are potentially hazardous to public health.

The checkpoint across the state border of the Russian Federation - the territory within the railway, automobile station, station, sea, river port, airport, airfield, open for international communications (international flights), as well as another specially equipped place where the border, and if necessary, and other types of control and passage through the state border of the Russian Federation of persons, vehicles, cargoes, goods and animals.

Back in the 14th century, Italy in Venice for the first time applied such a protective measure as the detention of ships, the cargo of people who arrived from disadvantaged areas of the world. The detention lasted 40 days. Hence comes the name quarantine. The success of this measure made it popular, and quarantines were built in the port cities of many European countries, buildings containing people who arrived from areas that were not at all suitable for the incidence of plague and cholera. In Russia, "border outposts" and "quarantines" appeared in the 16th century. In the future, these activities were improved and changed. However, over time, it became apparent that the quarantine system has become a means of economic and political influence of some countries on others.

In this regard, in 1969, the World Health Organization (WHO) introduced international health rules, and in 1981 some changes were introduced in connection with the elimination of smallpox. They change the basic concept of activity, which is now expressed in the sanitary protection of the territory, and not just the borders of the country.

Currently, the growing importance of epidemiological surveillance aimed at identifying and combating infectious diseases has been taken into account.

Sanitary rules establish the requirements for organizing, conducting and monitoring the implementation of measures aimed at preventing the importation and spread of quarantine diseases, contagious viral hemorrhagic virus fevers, malaria and other infectious diseases dangerous for humans transmitted by mosquitoes, localization and elimination of cases of these infections when they are imported or identification in the territory of the Russian Federation,

176

including in endemic natural foci, as well as the prevention of the import and distribution of goods potentially hazardous to public health. The purpose of international health regulations is to ensure that, without violating international transport and communications, maximum protection against the spread of diseases on an international scale. International health regulations apply to plague, cholera, and yellow fever. If these diseases occur, they notify WHO within 24 hours. In addition, information is provided on whether the detected cases of diseases are imported or of local origin.

The notification shall indicate the number of detected cases of quarantine diseases on board an arriving ship or aircraft. WHO concentrates these data and, in turn, regularly provides all countries with current and periodically reviewed epidemiological information on quarantine infectious diseases.

Sanitary protection of the territory of the Russian Federation is carried out by bodies and institutions of the State Sanitary and Epidemiological Service of Russia and is regulated by the law "On the Sanitary and Epidemiological Well-Being of the Population."

According to the rules, all medical institutions, regardless of their subordination and form of ownership, and those engaged in private medical practice, immediately (but no later than 24 hours) inform the centers of the State Sanitary and Epidemiological Surveillance about each case of a disease, suspected disease, and about the carriers of the pathogens listed above diseases, as well as cases of non-communicable diseases (poisoning) associated with imported food products or exposure to toxic substances.

5.1. Prevention of the plague importation.

The incubation period of the plague, in accordance with the current International Health Regulations, is set for a period of 6 days.

Vehicle owners must provide vehicles with everything necessary to prevent rodents from getting into and destroy rodents and ectoparasites if they are found on the vehicle.

The sea (river) vessels carry out deratization, and according to the testimony of the work done, they are granted the relevant documents. The list of sea and river ports entitled to issue a certificate of deratization,

177

established by the State Sanitary and Epidemiological Supervision of Russia and submitted to WHO. If rodents or traces of their vital activity are found as a result of the inspection of the vessel, the owner of the vessel is obliged to ensure deratization under the control of officials of the centers of Sanitary and Epidemiological Surveillance.

The deratization of the vessel is carried out with empty holds. After completion of the disinfestation, the centers of Sanitary and Epidemiological Surveillance issue a certificate of deratization. These measures may also apply to aircraft, rail and road transport. Certificates of exemption from deratization are valid for 6 months. When departing on an international flight from the country where cases of pulmonary plague are registered, each person taken aboard a vehicle is subject to medical supervision for 6 days.

A vessel, plane, train or road transport is considered infected upon arrival if there is a sick (suspected) plague on board.

Vehicles upon arrival are considered suspected of being infected with plague in the following cases:
- if there is no patient with plague, however, a case of this disease among passengers and crew occurred within the previous 6 days;
- detection of the death of rodents from an undetermined cause;
- detection of rodents on a vehicle emerging from a plague-enzootic area;
- if there is a person among the passengers or crew who is at risk of contracting pulmonary plague.

A sick (with suspicion) plague detected on a ship during a voyage is subject to immediate hospitalization at the ship's isolation ward before arrival at the nearest port, where there are necessary conditions for hospitalization.

Upon arrival of a suspect in the infection of vehicles in the Russian Federation, the following measures are taken:
-taking the vehicle to the sanitary berth;
-medical monitoring of passengers and crew for a period of 6 days from the moment of arrival;
- disinfection, disinsection and deratization in case of detection of rodents.

When a plague is detected on the territory of the Russian Federation by a citizen who has made an international flight or was located in a natural foci of plague within the country, they carry out all the necessary anti-epidemic and preventive measures in accordance with the requirements of these sanitary rules.

178

5.2. Prevention of cholera importation

Over 350,000 cholera diseases are registered annually in the world. The incubation period for cholera, in accordance with the current International Health Regulations, is set for a period of 5 days. If a patient is identified on a flight with signs among crew members and passengers following from infected areas, the vehicle administration is obliged to notify the sanitary authorities of the nearest port of the identified patient and to request the possibility of hospitalization.

When a patient with cholera on a vessel is detected during a voyage prior to arrival at the port of disembarkation, the patient is subject to hospitalization and immediate pathogenetic etiotropic therapy in a ship's isolation ward. Upon arrival at the port, the following measures are taken:

-taking the vehicle to the sanitary parking lot;
- mandatory hospitalization of the patient in compliance with the requirements of the epidemiological regime;
- isolation or medical monitoring of passengers for a period not exceeding the incubation period;
-bacteriological examination and preventive treatment of the Russian citizens, according to epidemiological indications;
- disinfection of the vehicle.

During the voyage and upon arrival at the port prior to clearance, it is prohibited to discharge overboard the patient's waste and their care items.

Food products, which are the cargo of the vehicle where a case of cholera during the trip was detected, if there is an epidemiological indication, and if the cargo is intended for sale on the territory of the Russian Federation, then they are subjected to bacteriological examination for the presence of cholera pathogens.

5.3. Requirements for the prevention of the yellow fever

importation.

The incubation period of yellow fever, in accordance with the current International Health Regulations, is set for a period of 6 days. When identifying a patient (with suspicion) with yellow fever on an arriving vehicle or during a flight, the latter is subject to hospitalization in an infectious diseases hospital (or cabin).

179

A valid yellow fever vaccination certificate is required for every member of a ship or aircraft crew using any port or airport located in an area infected with yellow fever.

Citizens of the Russian Federation traveling to the countries with areas infected with yellow fever are recommended to receive preventive vaccination. The absence of an international certificate of vaccination or revaccination against yellow fever by a citizen of the Russian Federation cannot be an obstacle to leaving the territory of the Russian Federation, but they are warned about the possibility of contracting yellow fever and the possibility of a delay at the airport of arrival. In the register of departing people, an entry is made about the refusal of vaccination and the warning with the signature of the departing citizen.

5.4. Special requirements for the prevention of the importation of malaria and other diseases dangerous to humans by mosquitoes.

Any vehicle resulting from areas endemic for malaria or other diseases transmitted by mosquitoes must be free from mosquitoes, for which it is subjected to disinfestation, information about which is entered in the marine sanitary declaration or in the sanitary part of the aircraft declaration. When mosquitoes are detected on such vehicles, at checkpoints across the state border of the Russian Federation, they are disinfected before unloading.

When identifying a patient with malaria during a flight from an endemic area, measures are taken to hospitalize him in the nearest port.

Chemoprophylaxis at ports of call should be carried out before and during the trip of a vessel calling at the port of endemic countries. When the sun is parked in the port, where there is a risk of contracting diseases transmitted by mosquitoes, captains (commanders) ensure that crews and passengers use repellents, especially when they are outdoors in the evening and at night, as well as the use of nets on doorways and windows.

180

REFERENCE LIST

Akatov A.K., Zueva V.S. Staphylococci. Moscow "Medizina", 1983.

Borovskaya M.F., Ordov V.P., Serko S.A. Veterinary-sanitary inspection. Staint-Petersburg. Moscow. Krasnodar, 2007.

Boyd U. Immunology basics. Moscow "Mir", 1969.

Gershanovich V.N. Biological chemistry and genetics of ion transport with bacteria. Moscow "Medicine", 1980.

Genetsianskaya T.A., Dobrovolskaya A.A. Special parazitology v.1-2. Moscow "Vyisshaya shkola", 1978.

Zharikova G.G. Food products microbiology. Sanitation and hygiene. Moscow Academia, 2005.

Kamysheva K.S. Microbiology, epidemiology basics and microbiology testing methods. Rostov-na-Donu "Feniks", 2010.

Krasilnikov N.A. Antagonism of microbes and antibiotic substances. Moscow "Sovetskaya nauka", 1958.

Kvasnikov E.I., Isakova D.M. Physiology of thermotolerant

microorganisms. Moscow "Nauka", 1978.

Kuvaeva I.B. Metabolism in organism and gut flora. Moscow "Medizina", 1976.

Marmuzova L.V.Basics of microbiology, sanitary and hygiene in food industry. Moscow, 2000.

Mikulenskiy S.R.History of biology. Moscow "Nauka", 1972.

Pozniakovskiy V.M. Hygienic basics of nutrition, safety and alimentary product inspection. Novosibirsk, 2002.

Pyatkin K.D., Kryvoshein J.S.Microbiology. Moscow "Medizina", 1976.

Fradkin Allergens. Moscow "Medizina", 1978.

181

CONTENTS

184

185

Microbiology and epidemiology
The textbook on microbiology, basics of epidemiology and microbiological research includes general and private microbiology, general epidemiology, the teaching about infectious and invasion diseases. The textbook systemically sets out the sections of microbiology according to the program approved by the Ministry of Health. The questions of morphology, classification, structure and nutrition of microorganisms, as well as issues of protein, carbohydrate, lipid and mineral metabolism are reviewed. The material on the study of the natural habitat of microorganisms is described in detail. The programs of infectious process occurrence and the dynamics of its spread are considered. The material on food toxicoinfections and toxicoses is set out detailly. The chapter on sanitary protection of the country territory from the import and spread of particularly dangerous diseases for a humann also occupies a prominent place. The textbook is intended for students and masters of the

Faculty of Veterinary Medicine, zootechnicians, medical professionals and students studying the technology of catering products manufacturing.

Georgii Kuzmich Vasiliadi, Doctor of Biological Sciences, Professor, senior scientist researcher, associate member of the Russian Academy of Natural Sciences. The author of 160 scientific works, including 4 books with the general circulation of 100 000 copies, 12 patents. Honoured Scientist of the Republic of North Ossetia-Alania, honoured worker of higher vocational education of the Russian Federation. Awarded with a golden medal at the VDNKh in 1970. Awarded with golden medal "To the glory of Ossetia".